DSL For Dummies, 2nd Edition

S0-BNB-883

Cheat Sheet

ILEC and CLEC Web Sites

ILEC/CLEC	Web Site
Ameritech (ILEC)	www.ameritech.com/products/data/adsl
Bell Atlantic (ILEC)	www.bell-atl.com/infospeed
Bell South (ILEC)	consumer.bellsouth.net/ adsldsl.smlbiz.bellsouth.com
Covad Communications (CLEC)	www.covad.com
GTE (ILEC)	www.getd.com/dsl
NorthPoint Communications (CLEC)	www.northpoint.net
Pacific Bell (ILEC)	www.pacbell.com
Rythms Net Connections (CLEC)	www.rhythms.net
Southwestern Bell (ILEC)	www.swbell.com/dsl
US WEST (ILEC)	www.uswest.com/products/data/dsl

DSL Flavors

DSL Flavor	What It Does
ADSL (Asymmetric DSL)	Delivers simultaneous high-speed data and POTS voice service over the same telephone line. Supports a range of speeds from 1.5 Mbps to 8 Mbps downstream and 1.54 Mbps upstream.
G.lite	Supports POTS and data on a single line. A variant of ADSL based on the G.lite standard that supports up to 1.5 Mbps downstream and 512 Kbps upstream, G.lite is intended for the DSL consumer and SOHO markets. This standards-based DSL is easier and less expensive for DSL providers to install and bring down the cost of DSL service.
SDSL (Symmetric DSL)	Supports symmetric service at 160 Kbps to 2.3 Mbps. SDSL does not support POTS connections but is suited for businesses that need the same high-speed data rates for upstream and downstream to support Internet servers.
IDSL (ISDN DSL)	Offers an always-on alternative to dial-up ISDN service with a capacity of 144 Kbps. Used where other forms of DSL cannot be delivered because the customer's premises are too far away from the central office.
HDSL (High-data-rate DSL) HDSL-2	Supports symmetric service at 1.54 Mbps but does not support POTS. HDSL is the DSL service widely used for T1 lines. HDSL uses four wires (two pairs) instead of the standard two wires used for other DSL flavors. A new version, HSDL-2, provides the same speed capabilities as HDSL but uses only a single-wire pair.
VDSL (Very-high-bit-rate DSL)	Supports up to 51 Mbps at very short distances. VDSL is the high-end member of the xDSL family.

For Dummies: *Bestselling Book Series for Beginners*

DSL For Dummies, 2nd Edition

Cheat Sheet

Questions to Ask an ISP about DSL Internet Service

- ✔ Do you offer price guarantees if DSL prices go down?
- ✔ Do you place usage restrictions on the amount of data going through the DSL connection?
- ✔ What are the terms of payment? Is there a time commitment with charges for early termination?
- ✔ What DSL CPE do you offer and support? What are the prices of the DSL CPE?
- ✔ Do you offer multiuser bridged and routed DSL service?
- ✔ Do you have any special deals on security solutions such as personal firewalls, proxy servers, or Internet security appliances?
- ✔ How many routable IP addresses are included as part of the DSL service? Can I get additional IP addresses? If so, how much do additional routable IP addresses cost per month?
- ✔ Do you restrict the operation of TCP/IP application servers on the DSL connection?
- ✔ Do you restrict the number of computers or users using the DSL service?
- ✔ How much do additional ISP-hosted email boxes cost per month?
- ✔ Does your DSL service use static or dynamic IP addressing?
- ✔ Do you offer customer support 24 hours a day, 7 days a week?
- ✔ Do you support DNS? If so, do you charge for domain name registration services?

DSL Shopping Checklist

- ✔ Start your DSL availability search at the ILEC and CLEC Web sites.
- ✔ Verify the DSL flavor being used by the DSL provider/ISP.
- ✔ Shop around to compare DSL offerings from all the ISPs in your area. Check the ILEC and CLEC Web sites for ISP partners offering DSL service in your area.
- ✔ Look beyond just the price of the DSL to include all the variables of the Internet access package.
- ✔ Know the minimum bandwidth you want and the upstream and downstream speeds.
- ✔ Understand your DSL CPE options (and their prices) as part of your DSL service.
- ✔ Read the fine print.
- ✔ Ask about any special promotions.
- ✔ Calculate the total cost of your DSL service, including all start-up costs and monthly fees.
- ✔ Determine whether you plan to run any TCP/IP application servers or connect multiple computers on your DSL connection.

DSL CPE Vendor Sites

Vendor	Product Information
Alcatel	www.usa.alcatel.com
3Com	www.3com.com
Cayman Systems	www.cayman.com
Cisco	www.cisco.com
Efficient Networks	www.efficient.com
Flowpoint	www.flowpoint.com
Netopia	www.netopia.com
Ramp Networks	www.ramp.com
Xpeed	www.xpeed.com

For Dummies®: Bestselling Book Series for Beginners

DSL

FOR

DUMMIES®

2ND EDITION

DSL
FOR
DUMMIES®
2ND EDITION

by David Angell

IDG Books Worldwide, Inc.
An International Data Group Company

Foster City, CA ◆ Chicago, IL ◆ Indianapolis, IN ◆ New York, NY

DSL For Dummies,® 2nd Edition

Published by
IDG Books Worldwide, Inc.
An International Data Group Company
919 E. Hillsdale Blvd.
Suite 400
Foster City, CA 94404
www.idgbooks.com (IDG Books Worldwide Web site)
www.dummies.com (Dummies Press Web site)

Library of Congress Control Number: 00-101254

ISBN: 0-7645-0715-X

Printed in the United States of America

10 9 8 7 6 5 4 3 2 1

2O/QS/QW/QQ/IN

Distributed in the United States by IDG Books Worldwide, Inc.

Distributed by CDG Books Canada Inc. for Canada; by Transworld Publishers Limited in the United Kingdom; by IDG Norge Books for Norway; by IDG Sweden Books for Sweden; by IDG Books Australia Publishing Corporation Pty. Ltd. for Australia and New Zealand; by TransQuest Publishers Pte Ltd. for Singapore, Malaysia, Thailand, Indonesia, and Hong Kong; by Gotop Information Inc. for Taiwan; by ICG Muse, Inc. for Japan; by Intersoft for South Africa; by Eyrolles for France; by International Thomson Publishing for Germany, Austria and Switzerland; by Distribuidora Cuspide for Argentina; by LR International for Brazil; by Galileo Libros for Chile; by Ediciones ZETA S.C.R. Ltda. for Peru; by WS Computer Publishing Corporation, Inc., for the Philippines; by Contemporanea de Ediciones for Venezuela; by Express Computer Distributors for the Caribbean and West Indies; by Micronesia Media Distributor, Inc. for Micronesia; by Chips Computadoras S.A. de C.V. for Mexico; by Editorial Norma de Panama S.A. for Panama; by American Bookshops for Finland.

For general information on IDG Books Worldwide's books in the U.S., please call our Consumer Customer Service department at 800-762-2974. For reseller information, including discounts and premium sales, please call our Reseller Customer Service department at 800-434-3422.

For information on where to purchase IDG Books Worldwide's books outside the U.S., please contact our International Sales department at 317-596-5530 or fax 317-572-4002.

For consumer information on foreign language translations, please contact our Customer Service department at 1-800-434-3422, fax 317-572-4002, or e-mail rights@idgbooks.com.

For information on licensing foreign or domestic rights, please phone +1-650-653-7098.

For sales inquiries and special prices for bulk quantities, please contact our Order Services department at 800-434-3422 or write to the address above.

For information on using IDG Books Worldwide's books in the classroom or for ordering examination copies, please contact our Educational Sales department at 800-434-2086 or fax 317-572-4005.

For press review copies, author interviews, or other publicity information, please contact our Public Relations department at 650-653-7000 or fax 650-653-7500.

For authorization to photocopy items for corporate, personal, or educational use, please contact Copyright Clearance Center, 222 Rosewood Drive, Danvers, MA 01923, or fax 978-750-4470.

is a registered trademark under exclusive license to IDG Books Worldwide, Inc. from International Data Group, Inc.

About the Author

David Angell is a recognized industry expert in DSL implementation for business and consumer communities. David is a 15-year computer and telecommunications industry veteran and the author of 20 books. He is the "DSL Evangelist" at a leading data CLEC, where he champions the needs of DSL subscribers and educates businesses and consumers about DSL. David lives in northern California. You can contact him at david@angell.com.

ABOUT IDG BOOKS WORLDWIDE

Welcome to the world of IDG Books Worldwide.

IDG Books Worldwide, Inc., is a subsidiary of International Data Group, the world's largest publisher of computer-related information and the leading global provider of information services on information technology. IDG was founded more than 30 years ago by Patrick J. McGovern and now employs more than 9,000 people worldwide. IDG publishes more than 290 computer publications in over 75 countries. More than 90 million people read one or more IDG publications each month.

Launched in 1990, IDG Books Worldwide is today the #1 publisher of best-selling computer books in the United States. We are proud to have received eight awards from the Computer Press Association in recognition of editorial excellence and three from Computer Currents' First Annual Readers' Choice Awards. Our best-selling ...For Dummies® series has more than 50 million copies in print with translations in 31 languages. IDG Books Worldwide, through a joint venture with IDG's Hi-Tech Beijing, became the first U.S. publisher to publish a computer book in the People's Republic of China. In record time, IDG Books Worldwide has become the first choice for millions of readers around the world who want to learn how to better manage their businesses.

Our mission is simple: Every one of our books is designed to bring extra value and skill-building instructions to the reader. Our books are written by experts who understand and care about our readers. The knowledge base of our editorial staff comes from years of experience in publishing, education, and journalism — experience we use to produce books to carry us into the new millennium. In short, we care about books, so we attract the best people. We devote special attention to details such as audience, interior design, use of icons, and illustrations. And because we use an efficient process of authoring, editing, and desktop publishing our books electronically, we can spend more time ensuring superior content and less time on the technicalities of making books.

You can count on our commitment to deliver high-quality books at competitive prices on topics you want to read about. At IDG Books Worldwide, we continue in the IDG tradition of delivering quality for more than 30 years. You'll find no better book on a subject than one from IDG Books Worldwide.

John J. Kilcullen
John Kilcullen
Chairman and CEO
IDG Books Worldwide, Inc.

Eighth Annual Computer Press Awards ≥1992

Ninth Annual Computer Press Awards ≥1993

Tenth Annual Computer Press Awards ≥1994

Eleventh Annual Computer Press Awards ≥1995

IDG is the world's leading IT media, research and exposition company. Founded in 1964, IDG had 1997 revenues of $2.05 billion and has more than 9,000 employees worldwide. IDG offers the widest range of media options that reach IT buyers in 75 countries representing 95% of worldwide IT spending. IDG's diverse product and services portfolio spans six key areas including print publishing, online publishing, expositions and conferences, market research, education and training, and global marketing services. More than 90 million people read one or more of IDG's 290 magazines and newspapers, including IDG's leading global brands — Computerworld, PC World, Network World, Macworld and the Channel World family of publications. IDG Books Worldwide is one of the fastest-growing computer book publishers in the world, with more than 700 titles in 36 languages. The "...For Dummies®" series alone has more than 50 million copies in print. IDG offers online users the largest network of technology-specific Web sites around the world through IDG.net (http://www.idg.net), which comprises more than 225 targeted Web sites in 55 countries worldwide. International Data Corporation (IDC) is the world's largest provider of information technology data, analysis and consulting, with research centers in over 41 countries and more than 400 research analysts worldwide. IDG World Expo is a leading producer of more than 168 globally branded conferences and expositions in 35 countries including E3 (Electronic Entertainment Expo), Macworld Expo, ComNet, Windows World Expo, ICE (Internet Commerce Expo), Agenda, DEMO, and Spotlight. IDG's training subsidiary, ExecuTrain, is the world's largest computer training company, with more than 230 locations worldwide and 785 training courses. IDG Marketing Services helps industry-leading IT companies build international brand recognition by developing global integrated marketing programs via IDG's print, online and exposition products worldwide. Further information about the company can be found at www.idg.com.   1/26/00

Dedication

This book is dedicated to the love of my life, my wife Joanne, for all her love, understanding, and encouragement.

Author's Acknowledgments

Although I'm the author of this book, it wouldn't have been possible without the help of many people along the way. A special thanks to Susan Pink, my project editor, for another outstanding job. Big kudos go to Dave Burstein, who is Editor of DSL Prime (www.dslprime.com), and Mark Peden, who is Director of Technology Standards, NorthPoint Communications and a member of the DSL Forum Board of Directors, for their bountiful insights as the technical editors for this book. Also a special thanks to Fred Balin at Mac Resolutions and Amy Peterson for rounding out the edits of this book. My thanks to the many people and companies in the DSL industry for providing essential support for this project.

Publisher's Acknowledgments

We're proud of this book; please register your comments through our IDG Books Worldwide Online Registration Form located at http://my2cents.dummies.com.

Some of the people who helped bring this book to market include the following:

Acquisitions, Editorial, and Media Development

Project Editor: Susan Pink

Acquisitions Editor: Georgette Blau

Technical Editors: Dave Burstein, Mark Peden, Amy Peterson

Senior Editor, Freelance: Constance Carlisle

Editorial Assistants: Beth Parlon and Candace Nicholson

Production

Project Coordinator: Emily Perkins

Layout and Graphics: Joe Bucki, Barry Offringa, Tracy K. Oliver Brent Savage, Jacque Schneider, Janet Seib, Brian Torwelle, Erin Zeltner

Proofreaders: Laura Albert, Vickie Broyles, John Greenough, Susan Moritz, York Production Services, Inc.

Indexer: York Production Services, Inc.

Special Help
Amanda M. Foxworth

General and Administrative

IDG Books Worldwide, Inc.: John Kilcullen, CEO

IDG Books Technology Publishing Group: Richard Swadley, Senior Vice President and Publisher; Walter R. Bruce III, Vice President and Publisher; Joseph Wikert, Vice President and Publisher; Mary Bednarek, Vice President and Director, Product Development; Andy Cummings, Publishing Director, General User Group; Mary C. Corder, Editorial Director; Barry Pruett, Publishing Director

IDG Books Consumer Publishing Group: Roland Elgey, Senior Vice President and Publisher; Kathleen A. Welton, Vice President and Publisher; Kevin Thornton, Acquisitions Manager; Kristin A. Cocks, Editorial Director

IDG Books Internet Publishing Group: Brenda McLaughlin, Senior Vice President and Publisher; Sofia Marchant, Online Marketing Manager

IDG Books Production for Branded Press: Debbie Stailey, Director of Production; Cindy L. Phipps, Manager of Project Coordination, Production Proofreading, and Indexing; Tony Augsburger, Manager of Prepress, Reprints, and Systems; Laura Carpenter, Production Control Manager; Shelley Lea, Supervisor of Graphics and Design; Debbie J. Gates, Production Systems Specialist; Robert Springer, Supervisor of Proofreading; Kathie Schutte, Senior Page Layout Supervisor; Michael Sullivan, Production Supervisor

Packaging and Book Design: Patty Page, Manager, Promotions Marketing

◆

The publisher would like to give special thanks to Patrick J. McGovern, without whom this book would not have been possible.

◆

Contents at a Glance

Cartoons at a Glance

By Rich Tennant

"THE IMAGE IS GETTING CLEARER, NOW... I CAN ALMOST SEE IT... YES! THERE IT IS — THE GLITCH IS IN A FAULTY CABLE AT YOUR OFFICE IN DENVER."

page 257

"Ooops — here's the problem. Something's causing shorts in the server."

page 173

BOB'S DECISION NOT TO BE CONNECTED TO THE COMPUTER NETWORK CAUSED SOME SUSPICION ON THE PART OF THOSE WHO WERE.

page 7

page 111

"BETTER CALL MIS AND TELL THEM ONE OF OUR NETWORKS HAS GONE BAD."

page 293

Fax: 978-546-7747
E-mail: richtennant@the5thwave.com
World Wide Web: www.the5thwave.com

Table of Contents

Introduction

· ·

Digital Subscriber Line, or simply DSL, is a high-speed Internet connection piped into a home or business through the humble telephone line. With DSL, the Internet springs to life. Without it, the full promise of the Internet is just that — a promise.

DSL delivers blazing speed Internet access of 10 to 100 times the speed of your dial-up modem. But wait, there's more! DSL is an always-on connection, which means you always have instant Internet access. No more busy signals, no dropped connections, and no time limits. How much does all this cost? You can tap into DSL power for as little as $39.95 a month (includes DSL and Internet access).

DSL breathes new life into those tired old telephone lines, transforming them into high-speed, digital Internet connections. At your home or business, a DSL modem connects to the telephone line and to your computer. At the other end of the line, DSL-enabling equipment is installed in the telephone company's central office (CO).

The Telecommunications Act of 1996 unleashed powerful competitive forces in the delivery of DSL. Today, DSL service is offered by not only the telephone companies but also numerous new competitive data communications companies, with names such as Covad, NorthPoint, and Rhythms. This unprecedented competition is driving the aggressive deployment of DSL to the point that in many metropolitan areas, several DSL providers are competing for your business. As a DSL customer, you're the winner with more choices and lower prices.

Millions of small businesses, telecommuters, and consumers now have the freedom to soar above the confines of slow, dial-up Internet access to change the way they live and work. DSL-powered Internet access affects your household or business the minute the service is turned on — and the uses and benefits of DSL service expand in ever-growing circles. At home, DSL integrates the Internet into your daily life. For businesses, DSL Internet access creates new opportunities to work smarter.

About This Book

DSL For Dummies is your springboard for diving into the brave new world of DSL connectivity to the Internet. This hands-on guide presents your DSL options in a real-world context, cutting through the hype and techno babble to give you the specific information you need to be an educated DSL customer. I explain the interrelated components that make up the entire ecosystem of DSL-based Internet access, including telecommunications, IP networking, and DSL hardware. Armed with this guide, you'll be able to pull together the optimal DSL Internet service for your needs.

Foolish Assumptions

Assumptions can get you into a lot of trouble, but in writing this book, I made some assumptions about who you are and what you're probably looking for from DSL service. This book is for you if

- You've had it with dial-up modems that have made your Internet experience painfully slow.

- You want to do more online from home, like shopping, investing, and telecommuting but it takes too much time using a slow modem connection.

- You want to send big files to other people connected to the Internet or download software from Internet sites without taking an hour or more.

- You've had it with high-priced ISDN from your local telephone company, and you're outgrowing its 128-Kbps limitation.

- You've been frustrated at the lack of affordable, high-speed data connections to link your business to the Internet the way the big guys with their T1 lines do.

- You've heard about DSL in some of its different incarnations (ADSL, SDSL, and so on), but you don't have a grasp of what DSL is all about and what it can do for you.

- You'd like to connect all the computers in your home or office to a single high-speed Internet connection.

- You want to know your options before you talk to an ISP about getting DSL service.

- You don't have a lot of extra time. You want to know only the amount of technical mumbo-jumbo necessary to make the right decisions.

- Your business or organization has a limited budget, but you still want people in your office to be more productive on the Internet.

✔ You're looking for a faster and more reliable way to telecommute to the office.

✔ You're using one or more computers using Microsoft Windows 95/98, Windows NT/2000, or the Mac OS and want to connect them to DSL service.

✔ You're on the lookout for new ways to gain a competitive advantage in today's business environment.

Conventions Used in This Book

This book's few conventions are shorthand ways to visually designate specific information from ordinary text. Here are the conventions:

✔ New terms are identified by using *italic*.

✔ Web site URLs (addresses) are designated by using a `monospace` font.

✔ Any command you enter at a command prompt is shown in bold and usually set on a separate line. Set-off text in italic represents a placeholder. For example, the text might read:

At the DOS prompt, enter the command in the following format:

ping *IPaddress*

where *IPaddress* is the IP address of the remote host you want to check.

✔ Command arrows, which are typeset as ⇨, are used in a list of menus and options. For example, Start⇨Find⇨Files means to choose the Start menu, then choose the Find menu, and then choose the Files option. The appearance of an underline means you can press Alt plus the underlined letter instead of clicking the option with the mouse.

✔ Commands are shown with a plus sign, such as Ctrl+C. This means to hold down the Ctrl key while you press C.

How This Book Is Organized

DSL For Dummies has five parts that present material in a steady progression of what you need to know at the time you need to know it. Each part is self-contained but also interconnected to every other part.

Part I: Getting Comfortable with DSL

Part I lays the foundation for your journey to DSL enlightenment. You discover how DSL service works and get help finding your way through the tangle of competing DSL technologies. After getting the DSL fundamentals, you dive into how and where to shop for DSL Internet service. You find out how to check if DSL is available in your area and who offers it. I guide you through what you need to consider for the Internet access part of your DSL service. Finally, I pull it all together to give you a game plan for becoming an educated consumer of DSL service so you find the right DSL Internet package for your needs.

Part II: Making DSL Come Alive with Internet Access

A DSL connection is lifeless without Internet access. Part II covers the essential elements you need to get the most out of your DSL Internet service. You get help in understanding the fundamentals of TCP/IP (Transmission Control Protocol/Internet Protocol) as a foundation for choosing the right IP network configuration for your business or home. You get the lowdown on what IP services, such as IP addresses, domain name services, Web hosting, and e-mail hosting, to consider as part of your Internet connection.

You explore your DSL CPE (Customer Premises Equipment) options and what the differences between a DSL modem and router mean to you. After mastering TCP/IP and DSL CPE essentials, you move into the realm of Internet security risks and solutions. You see how you can share a DSL Internet connection where the ISP supplies only one IP address.

Part III: DSL Internet Service Meets Your Computer or LAN

Part III describes the mechanics of where the DSL connection meets your computer or LAN. If you want to connect multiple PCs or Macs to your DSL connection, you can find out how set up a network. I tell you how to set up the basic LAN plumbing for the three leading network types, including Ethernet, Phoneline, and wireless. Once you have a LAN in place, you have the foundation for sharing your DSL Internet service.

You also see how to configure Microsoft Windows 95/98, Windows NT/2000, or the Mac OS for different types of DSL CPE and Internet connections. If you're using an Ethernet DSL modem or router to connect to the Internet, you find out how to configure the TCP/IP settings for your NIC (network interface card). I also show you how to work with Microsoft Windows Dial-Up Networking to support certain types of DSL Internet service.

Part IV: Got DSL, Now What?

Part IV goes beyond the mechanics of DSL and gets into the good stuff. What can you do with a DSL Internet connection? Here I take you through the highlights of what broadband Internet access means to your household or business. You take a look at how broadband integrates the Internet into your daily life at home. Next, you see how the DSL powered business works smarter by tapping into online business-to-business services on the Internet. I show you how to add video conferencing and voice to your DSL connection to add show-and-tell to your Internet communications. You find out how DSL combined with virtual private networking (VPN) puts telecommuters in the fast lane.

Part V: The Part of Tens

The Part of Tens contains valuable resource nuggets packaged in groups of ten to help you on your way toward DSL enlightenment. Much of this information enhances and supports topics covered in previous chapters. You find helpful tips on how to shop for the best DSL service, and what questions to ask about DSL service.

Glossary

I include a comprehensive glossary to help you speak the language of DSL and broadband Internet.

What You're Not to Read

There is a lot of stuff you're not going to read in this book. You're not going to get into a technical jungle that goes off into splitting hairs. This book is about educating you on just what you need to get results — that is, the right DSL service and the right price. All technical discussions in this book play a supporting (not primary) role in helping you to make the right decisions.

Icons in This Book

DSL For Dummies includes icons that act as markers for special information. Here are the icons I use and what they mean.

This icon signals nerdy techno-facts that you can easily skip without hurting your TCP/IP education. But if you're a technoid, you probably eat this stuff up. Enrich your mind or skip 'em if you like.

This icon indicates nifty shortcuts or pieces of information that make your life easier. Here are the tips and tricks that will save you time, money, and aggravation.

This icon lets you know that there's a loaded gun pointed directly at your foot. Watch out! Watch out! DSL may be the fast lane for Internet connections, but the road has potholes and hairpin curves. This icon points out these problems and tells you how to avoid or solve them.

This icon flags the presence of Web-based resources that you can check out to get more information on DSL products, services, and other resources.

Where to Go from Here

You are about to make your first move to the fast lane of Internet connectivity. Crack this book open and dive into Chapter 1 to get an executive summary of DSL and what it can do for you. From there, feel free to move about the book in a systematic linear manner or in random acts of information gathering. Either way, *DSL For Dummies* guides you through the world of high-speed, always-on DSL connections.

Part I
Getting Comfortable with DSL

The 5th Wave   By Rich Tennant

BOB'S DECISION NOT TO BE CONNECTED TO THE COMPUTER NETWORK CAUSED SOME SUSPICION ON THE PART OF THOSE WHO WERE.

In this part . . .

Today's modem-based Internet access is like being stuck in rush-hour traffic. You're stranded in the slow lane, and fast and affordable access to the Internet seems like a pipe dream. Enter Digital Subscriber Line (DSL), your ticket to the fast lane.

In Part I, you dive into the brave new world of DSL and what it can do for you. You get a solid grounding in how DSL service works and get help navigating through the maze of competing DSL technologies.

Next, you take the important step of acquiring some DSL data communication fundamentals. Then you move on to checking out DSL providers to find out whether DSL service is available in your area.

After you determine DSL availability, I guide you through what you need to consider for the Internet access part of your DSL service. Finally, I pull it all together to give you a game plan for becoming an educated consumer of DSL service so you'll find the right DSL Internet package for your needs.

Chapter 1

DSL: More Than a Pipe Dream

• •

• •

*F*or the millions of businesses and individuals stuck in the slow lane of analog modems or ISDN, fast and affordable access to the Internet has been a pipe dream. Welcome to the frustrated world of the bandwidth have-nots.

Enter *Digital Subscriber Line (DSL),* a new data communications technology that is making high-speed bandwidth a reality for the rest of us. Why is DSL so promising? This powerful data communications technology works over regular telephone lines and is being delivered to homes and businesses in a revolutionary way: competitively. The telephone company is no longer the exclusive provider of data communications. You get to decide who delivers your DSL Internet access.

A Bad Case of the Bandwidth Blues

Bandwidth is the mantra of the Internet. Simply put, *bandwidth* is the capacity of any data communications link. The Web — today's *graphical user interface,* or *GUI,* to the Internet — and the explosion of multimedia technologies have created a bandwidth-hungry environment that overwhelms today's modems. Anyone clunking around the Internet by using a dial-up modem over a system designed for voice more than a hundred years ago has felt the bandwidth blues.

While the Internet's evolution continues to speed along, the connectivity options for the majority of businesses, telecommuters, and consumers have languished. Until the advent of DSL, the bandwidth options offered by the telephone companies couldn't help but give you a bad case of the bandwidth blues. On the low end of the bandwidth spectrum are dial-up modems and ISDN; on the high end are the dedicated, leased line services of frame relay and T1 services.

The good news is that telephone companies are responding to competition from a new breed of telecommunication companies offering DSL service. Many telephone companies are aggressively embracing DSL as a high-speed Internet service for the rest of us, thanks to the push of competition.

Slow-motion Internet: POTS and modems

For most Internet users, the analog modem is the current staple for data communications. A modem connection uses the same *POTS (Plain Old Telephone Service)* lines used for voice communications. This ubiquitous telephone system was designed to transmit the human voice as an analog waveform. Analog modems transmit digital computer information first by converting the digital data to analog signals. At the receiving end of the connection, a modem changes the analog signals back into the digital form used by computers.

The telephone lines from your home or office connect to a vast global telecommunications network referred to as the *Public Switched Telephone Network (PSTN)*. This network uses a complex system of computers to route calls based on telephone numbers. Although this system's infrastructure has been continually upgraded and converted to digital, the basic copper wiring between your home or business and the telephone network has been around for a long time. Analog-based data communications have inherent limitations that make today's 56 Kbps modems the end of the line for this technology.

For businesses, modem service has been no bargain. Unlike most residential customers, businesses typically pay usage charges for all local calls. Therefore, a small office with a few computers connected to the Internet using modems for a few hours a day can easily rack up hundreds of dollars a month in telephone company charges. Internet access charges from the Internet service provider (ISP) are added on top of these telephone costs.

Residential customer typically don't incur usage charges for local calls to the ISP but many use a second telephone line for Internet access, incurring the monthly cost of the second line plus the monthly ISP charges.

ISDN: Too little, too late, too expensive?

ISDN (Integrated Services Digital Network) was the first attempt by telephone companies to offer a higher speed, digital service to the mass market. ISDN delivers data at up to 128 Kbps over two 2 B (bearer) channels (at 64 Kbps each) using the same telephone wires used for POTS. The telephone companies have been deploying ISDN for several years, and it is widely available.

The telephone companies deploy ISDN as part of the PSTN, which makes most ISDN service a dial-up and metered service (although a few offer a more cost-effective unlimited rate ISDN services). Because ISDN is a premium service based on usage, it's expensive and impractical for Internet connectivity. Business ISDN service typically costs $100 to $350 for installation and $50 to $330 (or higher) per month. Residential ISDN service is no bargain in many areas. For example, in the Boston area, Bell Atlantic charges about $50 a month for just the line and 1¢ per minute per B channel for a local data call. The usage rate might seem nominal, but it adds up quickly when you're using both B channels for several hours a day.

If you're already using ISDN for your Internet access for more than 20 hours a month, you're a prime candidate for DSL, if it's available in your area. In most cases, moving up to DSL will not only dramatically speed up your Internet service but also save you money. In addition, switching from ISDN to DSL results in a dedicated connection to the Internet, which means no more dial-up process.

T1 and frame relay: High speed, high cost

The telephone company offers businesses *leased lines,* which are dedicated, high-speed links to the Internet installed between two points (called *point-to-point service*) to provide a dedicated (always-on) service. Users connecting to the Internet with a dedicated connection don't go through a dial-up process. Instead, connections are always ready and waiting for instant access. The leading types of leased and network services offered by the telephone companies are T1, fractional T1, and frame relay.

Although the cost of dedicated service has dropped, it remains prohibitively expensive for most individuals and small to medium-sized businesses. The local telephone companies continue to extract high prices for leased line services by using elaborate, outdated pricing models that have created a highly profitable business for them. T1 service typically includes Quality of Service (QoS) and quick repair guarantees, which telephone companies claim justifies the higher cost.

T1 lines are dedicated, leased lines that deliver data at 1.54 Mbps. The 1.54-Mbps pipeline is divided into 24 64-Kbps channels. These high-speed, digital links are created by using telephone lines with equipment to handle higher data rates. The cost of a T1 line varies from $500 to $1100 for installation and $450 to $750 (or higher) per month. At least two charges are associated with T1 service for Internet access: one from the local telephone company, and the other from the Internet service provider.

A more powerful version of the T1 line is the T3 line, which is the equivalent of 28 T1 lines. Large companies and ISPs use this service for their backbone networks. A *backbone* net is a high-volume data communications link that carries data consolidated from smaller data communication links.

Ironically, the telephone companies have used DSL technology in the form of HDSL *(High-bit-rate Digital Subscriber Line)* for years to offer T1 line service. A new standards-based version of HDSL, called HDSL-2, operates on only one pair of wires instead of the two required for HDSL. In addition, HDSL-2 minimizes interference when the wires run side-by-side with other data communication technologies. *Fractional T1* service is T1 service offered in fractional amounts based on 64-Kbps increments, such as 256 Kbps, 384 Kbps, and 784 Kbps. A customer can start with fractional T1 service at one capacity level and upgrade as demand warrants. When a customer orders fractional T1 service, the carrier sets up a full T1 line but makes only the contracted bandwidth available. Fractional T1 service is costly but less expensive than full T1 service.

Frame relay — a switched public network offering from the telephone companies — is another popular form of service. It's commonly delivered over T1 or fractional T1 lines. Frame relay is a data network service bundled with leased line access for transmitting data between remote networks. It's a network service built by local and long-distance telephone companies that acts like a private dedicated network. A typical 384-Kbps frame relay connection has installation costs from $700 to $1200 and a monthly cost from $550 to $850.

DSL: The Bandwidth Sweet Spot

Digital Subscriber Line (DSL) shares the same high-speed, always-on qualities of dedicated leased lines but at dramatically lower prices. DSL is actually a term used for a family of related telecommunications technologies that include such members as ADSL, SDSL, IDSL, and VDSL. Each of these DSL flavors has unique bandwidth capabilities and limitations.

DSL service is so promising because it uses the same copper wiring as regular voice telephone service. DSL equipment at each end of the line transforms the telephone line into a high-speed digital connection. The *DSL provider* is the telecommunications company that supplies the DSL portion of your high-speed Internet access. The *Internet service provider* (ISP) adds the Internet Protocol (IP) networking services that ride over the DSL connection.

The suite of DSL offerings is filling the chasm between slower dial-up modem and ISDN services and expensive T1 and frame relay services. DSL service is squarely targeted at this bandwidth sweet spot and goes beyond it to offer even more power than T1 service; for example, ADSL supports data speeds up to 8 Mbps.

DSL brings a new level of Internet connection to businesses, telecommuters, and consumers with the following powerful data communication benefits:

- **Always-on connectivity.** There is no dial-up process, and the Internet is available for two-way data communications 24 hours a day, 7 days a week (commonly referred to as a 24x7 connection). These same attributes are inherent in traditional leased line offerings. An always-on connection translates to new ways to use the Internet for businesses and consumers.

- **High speed.** DSL comes in a variety of speeds from 144 Kbps up to 8 Mbps. These bandwidth capabilities provide productive access to the full range of interactive content that the Internet has to offer. Most DSL service offerings also support bandwidth *scalability,* which means that you can increase your speed over time without incurring new start-up costs or buying new equipment.

- **Flat rate service.** The DSL link typically doesn't have usage-sensitive pricing, which means you can use the DSL connection any time for as long as you need without incurring usage charges. The Internet access component of your DSL service, however, might have usage-based pricing after a certain threshold, depending of the service offering.

- **Reliability.** Because DSL is pure digital communications, it's inherently a more reliable medium for data than using an analog modem over a voice grade telephone line.

- **Connection to the future.** High-speed, always-on DSL Internet connectivity enables your home or business to have a front row seat to all the Internet has to offer.

DSL is broadband

Technically speaking, *broadband* is any high-bandwidth communications technology (more than 200 Kbps) that allows the combined transmission of data, voice, and video streams within channels over a single physical connection. In the popular lexicon of the Internet, however, *broadband* describes any Internet connection that offers a fatter data pipe into your home or business than narrowband (dial-up analog modem service). Because most DSL flavors support over 200 Kbps, they fit the broadband definition.

A little DSL history

Digital Subscriber Line emanated from Bellcore, the research organization for the telephone companies formed after the breakup of AT&T. Telephone companies envisioned DSL as a way to deliver video services to compete with the cable TV industry. These services never materialized in the United States, and DSL languished until bandwidth demand generated by the Internet explosion gave DSL a new lease on life.

In the hands of the telephone companies operating in their traditional monopolistic environment, DSL might have become another high-priced service outside the reach of all but the largest organizations. But change is swirling around the telecommunications industry, unleashing powerful competitive forces. DSL service deployment is at the forefront of taking advantage of this new competitive environment.

The Telecommunications Act brings DSL to life

The new competitive environment driving the deployment of DSL was started by the passage of the Telecommunications Act in 1996. This legislation is opening up the local telecommunications industry to competition, which ultimately gives consumers more choices at lower costs. Although by no means a perfect piece of legislation, it's already bearing a bountiful harvest of DSL service options for businesses and consumers.

Local telephone companies, called *Incumbent Local Exchange Carriers (ILECs)*, control the local telephone service infrastructure. The interconnection provisions of the Telecommunications Act require ILECs to allow their services and network elements to be resold by competitors. These competitors are called *Competitive Local Exchange Carriers (CLECs)*. CLECs offering DSL data communication services exclusively are commonly referred to as *data CLECs,* or simply *DLECs*.

CLECs make DSL service available to a given geographical area by installing equipment at the ILECs' facilities, called *central* offices (COs). COs are the terminus points on the telephone company side for the copper lines that reach out to every home and business in the United States. CLECs are also adding special DSL-enabling equipment called MTUs (Multi-Tenant Units) inside larger residential and office buildings. These MTUs may bypass the CO to provide DSL service to everyone in the building.

Competitive pressure on ILECs is forcing them to move more quickly and efficiently in deploying their own DSL services. Competition, however, isn't coming from only CLECs. The cable industry is also deploying high-speed Internet access services through their networks at competitive prices.

CLECs operate in dangerous waters because their main competitors for DSL service (the ILECs) are also their main suppliers of two essential pieces of the DSL network: the telephone lines going to every home and business and space at the CO to install DSL enabling equipment. For CLECs, this blurred supplier/competitor relationship requires skillful management within a complex matrix of technical, financial, and regulatory factors. CLECs spend a lot of money to keep ILECs on the path of opening up their networks and ensuring competitive pricing parity. To this end, CLECs directly negotiate with ILECs or take issues to the doorstep of the FCC, state regulatory agencies, or the courts.

The bottom line on the Telecommunications Act is that it has unleashed the power of competition into the deployment of DSL service by CLECs and ILECs. Both DSL providers are aggressively building their DSL networks to make the service available to an ever-growing number of areas. DSL service deployment has gone beyond the large metropolitan areas and is now being deployed in many smaller cities and towns and even rural areas.

See Chapter 4 for specific DSL deployment by CLECs and ILECs and how you can check to see whether DSL service is available in your area.

Sharing is even better

A Federal Communications Commission (FCC) ruling in late 1999 continues to unleash the power of competition in the delivery of DSL service. The FCC ruled that ILECs must give CLECs the capability to offer ADSL and G.lite service over their existing voice telephone lines. The impact of this ruling, which takes effect in the second half of 2000, promises to give DSL consumers choices in DSL service providers, resulting in lower prices while accelerating service deployments.

DSL is an international phenomenon

The telecommunications industry is global, and DSL is being rolled out internationally. Canada and the U.K. are heavily wired for DSL and about to roll out service; Korea and Germany are also rolling out millions of lines. You can expect to see DSL continue its global reach in 2001.

Meet the DSL family

Bellcore developed Digital Subscriber Line technology as a technique to filter out the background noise, or interference, on copper wires to allow clearer connections. This filtering process enables more data to move through regular telephone lines. *DSL* refers not to the telephone line itself but to the technology that allows a high-frequency digital signal (hence more data) to pass over copper telephone wiring.

The generic term for the family of DSL services is xDSL (or often, simply DSL), where the *x* is a placeholder for the letter used for a specific flavor of DSL. Each type of DSL service has its own limitations and rate of data communications. Exact DSL service performance is based on the following factors:

- ✔ **The distance between a user's premises and the telephone company's central office**
- ✔ **The DSL equipment used at both ends of the connection**
- ✔ **The service offerings from ILECs and CLECs**

The distance of your premises from the CO is one of the most significant factors in determining the availability of DSL service and the type or speed of services offered. Many DSL flavors have distance ranges up to 18,000 feet from the CO; others go up to 36,000 feet. As the distance between a customer's premises and the CO increases, the maximum speed available is reduced. However, all these distance ranges are theoretical and based on the distance of the actual telephone wire — not the distance of the customer's premises from the CO. Telephone wiring can take a meandering path that ends up exceeding the distance range even though the home or business is within the range. In addition, other factors affect the actual speed to a specific home or business.

Table 1-1 describes the key attributes of the DSL family of technologies. The leading DSL services being deployed and targeted at the bandwidth sweet spot are ADSL, G.lite, SDSL, and IDSL. For small businesses and homes, ADSL and SDSL more widely deployed today. Chapter 3 goes into more specifics about each form of DSL.

Table 1-1	The xDSL Family of High Bandwidth Services
DSL Flavor	*What It Does*
ADSL (Asymmetric DSL)	Commonly called full-rate ADSL. Delivers simultaneous high-speed data and POTS voice service over the same telephone line. The latest ADSL implementations are based on the ITU-T G.dmt standards. Supports speeds of up to 8 Mbps downstream and up to 1.5 Mbps upstream.

DSL Flavor	What It Does
G.lite	A new variant of ADSL formerly called ADSL Lite or Universal ADSL. It's based on the ITU-T G.lite standard. It supports 1.5 Mbps downstream and 512 Kbps upstream. G.lite is intended for the mass market, including consumers, small businesses, and remote offices. G.lite, like G.dmt ADSL, delivers simultaneous high-speed data and voice service over the same telephone line.
SDSL (Symmetric DSL)	Supports symmetric service at 160 Kbps to 2.3 Mbps. SDSL does not support POTS connections. Most of the national CLECs offer SDSL targeted at small businesses and SOHOs.
IDSL (ISDN DSL)	Offers an always-on alternative to dial-up ISDN service with a capacity up to 144 Kbps. IDSL is often used where other forms of DSL cannot be delivered because the customer premises is too far away from the telephone company's central office to support ADSL, G.lite, or SDSL.
HDSL (High-bit-rate DSL), HDSL-2	Supports symmetric service at 1.54 Mbps but does not support concurrent POTS. HDSL is the DSL technology often used as the basis for newer T1 lines from the telephone companies. HDSL uses four wires (two pairs) instead of the standard two wires used for other DSL flavors. HSDL-2, provides the same speed capabilities as HDSL but uses only a single wire pair.
VDSL (Very high-bit-rate DSL)	Supports up to 51 Mbps at very short distances of between 1,000 to 4,000 feet. Like ADSL and G.lite, it can be delivered concurrently over a POTS line.

Upstream and downstream speeds

When people talk about data rates and DSL, they use the terms upstream and downstream. The *upstream data rate* refers to the speed data travels from your local computer or network to the Internet or other remote network. The *downstream data rate* refers to the speed data travels from the Internet or other remote network to your local computer or network.

Asymmetric DSL means that the downstream speed is higher than the upstream speed. For example, an ADSL connection might have a downstream speed of 1.5 Mbps and an upstream speed maximum of 512 Kbps. *Symmetric DSL* means that the downstream and upstream speeds are the same. DSL service rates are typically represented as downstream/upstream. For example, an asymmetric offering might be noted as 1.5 Mbps/512 Kbps, and a symmetric offering might be noted as 384 Kbps/384 Kbps service.

Delivering DSL to your home or business

DSL uses telephone lines to deliver high-speed, always-on data communications. What transforms these ordinary telephone lines into high-speed DSL lines is the special equipment at each end of the telephone line. At the home or business is a DSL modem or router that connects to the telephone line and to a computer or local area network (LAN). At the central office (CO) servicing your area, the telephone line terminates at special DSL-enabling equipment called DSL Access Multiplexers (DSLAMs). The DSLAM can also be installed in a large office or residential building instead of at the CO as an MTU.

The data from your PC travels through the DSL modem or router, out to the telephone line, and to the CO, where its routed to the DSLAM and then on to the DSL provider's data network to be passed off to your ISP.

If you're using ADSL or G.lite, your DSL service typically runs over the same telephone line as your voice service. The data service is split off from the voice service at the CO. Your voice communications goes to the telephone network and the data goes to the DSLAM of the DSL provider and on to the ISP, as shown in Figure 1-1.

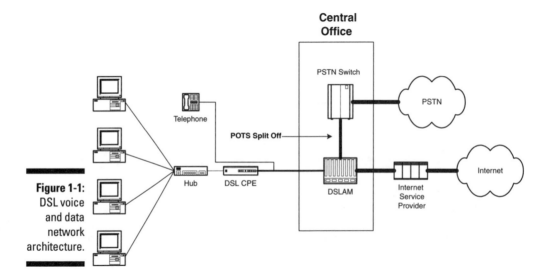

Figure 1-1:
DSL voice and data network architecture.

Don't get your circuits crossed

A *circuit* is a path through a network from one point to another. Communications over the Public Switched Telephone Network (PSTN) are handled by circuit switches, which route voice and data (dial-up analog modem and ISDN service) traffic from one destination to another. In a *circuit-switched network,* a temporary circuit, or path, is established between two points by dialing. The circuit is terminated when one end of the connection hangs up.

DSL service lines bypass the circuit-switching infrastructure of the PSTN and terminate at the CO, where they are routed to their own data network. DSL service uses a dedicated circuit that is always on, which means the path between your home or business to the Internet is a dedicated connection. In telecommunications circles, this type of data connection is commonly referred to as *nailed up.*

Some DSL flavors, such as SDSL service, require a separate line from your voice telephone line. The DSL traffic goes to the DSLAM, and voice communications over a voice telephone line go to the telephone network, as shown in Figure 1-2.

A new technology for DSL called *VoDSL* (Voice over DSL) will allow voice communications to pass over DSL as IP data packets and then pass off to the voice telephone network at the CO. Expect to see deployment of VoDSL service offerings in late 2000 or early 2001.

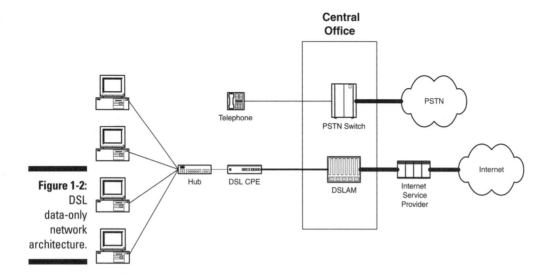

Figure 1-2:
DSL
data-only
network
architecture.

DSL is a bundled-meal deal

DSL is typically sold as part of a complete Internet access package. CLECs and ILECs can sell DSL service directly to consumers or wholesale to Internet service providers, who in turn package the DSL service with Internet access services to consumers and businesses. Most ILECs operate their own Internet service provider companies as well as sell to other Internet service providers. Likewise, some CLECs also sell DSL service both wholesale and direct.

DSL Networks Are Growing Like Kudzu

Deployment by competing DSL providers is spreading DSL availability like kudzu is taking over the southern part of the United States. Both ILECs and CLECs continue to aggressively roll out DSL services. And in many markets, you can choose between multiple DSL providers. In 2000, DSL deployment will rapidly expand into many smaller metropolitan areas and even to rural areas.

DSL deployment is on a CO-by-CO basis, which can result in a patchwork of service availability by some DSL providers looking to grab the most concentrated markets. The leading CLECs, such as Covad and NorthPoint, generally serve all the COs in a region, not just the most lucrative.

Because DSL comes in many flavors, the type of DSL available to you might be SDSL, ADSL, G.lite, or IDSL. Until recently, CLECs focused on the business market with SDSL service, and ILECs focused on ADSL for consumers. Now, however, both CLECs and ILECs are blurring this early DSL deployment distinction. With the emergence of line sharing and the introduction of standards-based DSL such as G.dmt and G.lite, you can expect CLECs to be aggressively offering DSL service to the home consumer.

What Will DSL Internet Service Cost?

Like DSL service options, pricing for DSL service is all over the map. The cost of DSL service depends on the type of DSL service available, the DSL provider, and the Internet service provider. DSL service is sold through ISPs, and as such they typically price the service within a given area. Their wholesale costs for the service depends on the DSL provider and the type of DSL service.

DSL service typically has a one-time start-up cost, which may include a one-time charge for the DSL service, the Internet access account, and a DSL modem or router. These one-time charges can range anywhere from free to over several hundred dollars, depending on the ISP, your location, the type of service, the equipment, and other factors. However, many ISPs offer special DSL service promotions (often funded by the DSL provider) that waive installation charges, provide a free DSL modem, and give the first month free in exchange for a service contract of a year or more.

You can't compare one DSL service offering to another on price alone because a number of factors go into the total DSL Internet service package, including the

- ✔ Speed of the connection
- ✔ DSL provider
- ✔ Type of DSL
- ✔ Bundled ISP services, such as IP addresses and email boxes
- ✔ Type of DSL CPE (consumer premises equipment)

Most DSL service, however, is broken down into two distinct price points depending on the intended user: business or consumer.

Business-class DSL is more expensive but also delivers a more reliable service and includes features that benefit businesses, such as the use of a router and provision for multiple IP addresses. The price point for this service ranges from $70 to $400 per month, depending on the speed and the Internet services.

Consumer DSL service is usually an asymmetrical offering (ADSL or G.lite) that supports more downstream speed than upstream speed. ADSL and G.lite are the preferred types of DSL for a residential consumer because they can be delivered over the same line as the existing voice (POTS) service.

DSL service offerings for the consumer market currently fall within the price range of $30 to $60 but will likely get down to $20 to $40. The basic entry-level DSL offering is an ADSL package with speeds such as 384/128 Kbps or 1.5 Mbps/384 Kbps.

Consumer Internet service typically provides a single IP address. As competition heats up in the consumer market utilizing G.lite and taking advantage of the FCC's line-sharing ruling, you can expect choices for consumers to expand and prices to fall.

The distinction between business and consumer markets is by no means hard and fast. Some telecommuters, home-based businesses, and professionals working out of a home office want the features associated with business class service. You must evaluate the attributes of each type of service to determine what works best for you.

Free DSL: You get what you pay for

Broadband Digital Group (BDG) is offering its FreeDSL service in 40 markets by year-end 2000. To subscribe to FreeDSL, consumers go to `www.freedsl.com`, download the user software, and open an account. The basic FreeDSL service offers speeds up to 144 Kbps. For $10 a month, you can get the same service as the basic FreeDSL without the desktop advertisements. BDG offers additional upgrade packages that increase speeds to 384 Kbps for $20 per month or to 1.54 Mbps for $35 a month.

You're required to buy a start-up kit for $199, which includes the DSL modem. The kit can be paid off in 23 monthly payments of $9.95 a month, for a total cost of $228.85. You can get the start-up kit for free if you refer ten qualified customers who sign up for FreeDSL.

The important thing to keep in mind about this "free" service is that it comes with strings attached. Read the terms carefully. Although you can save money on your DSL Internet connection, you may be giving up privacy and be required to view and click on advertisements to acknowledge you've viewed them.

A DSL Internet Connection Changes Everything

Like expensive dedicated services, DSL is an always-on connection, which means that you're open for business 24 hours a day, 7 days a week. The always-on, high-speed capabilities of DSL enable you to do many things, including the following:

- **Add show-and-tell to your Internet communications.** In the world of low bandwidth and dial-up modems, using IP voice and video conferencing simply wasn't a viable communications solution. The high-speed and always-on capabilities of DSL make IP voice and video conferencing a more practical, everyday tool. Although video conferencing doesn't have the video quality of television, it does open up visual communications over the Internet.

- **Cruise the Web with the top down.** At a very basic level, Internet connections through DSL become a thrill as your fingers do the traveling. High-speed, instant Internet access means consumers can use the Internet for a growing number of household tasks — such as shopping, banking, bill paying, investing, customer support, and learning. Businesses will find that their employees are more productive when working on the Web.

The always-on DSL connection means that double-clicking the Microsoft Internet Explorer or Netscape Communicator icon instantly delivers the Internet to your desktop. No more squealing modem connections that seem to take forever. No waiting for Web page downloads. In fact, your connection becomes so fast, you begin to see how slow different Web servers are in delivering information to you.

✔ **Create your own virtual organization.** DSL enables every participant in a virtual enterprise to be both a user and a provider of information. You can link up geographically dispersed groups to operate as a virtual team. Electronic whiteboards combined with video conferencing over a DSL connection can create a powerful set of real-time, interactive tools that bring dispersed teams of people to new levels of communication.

✔ **Push and pull your applications.** The Internet is rapidly becoming the software distribution medium of choice. Entire programs can be quickly downloaded from a software publisher's Web site. Using push and pull technologies, software vendors can even send updates to your software automatically. Push technologies can deliver customized news directly to any desktop on your LAN (local area network). News, stock quotes, and other information can be constantly streaming down to your business.

✔ **Move your content rapidly.** As computer applications become more sophisticated, the size of programs and files grows almost exponentially. Desktop publishers, multimedia developers, software publishers, lawyers, architects, and a host of others can deliver digital content economically through DSL connections either as a download or by providing file servers. Downloading a 10MB file, for example, takes 32 minutes using a 56 Kbps (56K modems have an average speed of 42 Kbps). Using a 784-Kbps DSL connection, downloading a 10MB file takes less than 2 minutes.

✔ **Telecommute in the fast lane.** DSL is a powerful tool for making you feel like you're sitting at a desk in the corporate office even though you're working remotely. Using *virtual private networking (VPN)* over the Internet, telecommuters can do work on the Internet as well as link up to computers on a company network. DSL-based remote access enables sophisticated and bandwidth-hungry applications, such as video conferencing and Web-based collaborative computing, to become everyday telecommuter tools.

✔ **Send and receive email instantly**. With always-on Internet access, you can get email messages the instant they arrive on the ISP's email server. Likewise, your email is sent to the recipient the instant you click the send button. The high speed of a DSL Internet connection also lets you quickly send large file attachments as email attachments.

If you use DSL, you're a subscriber

In the lexicon of telecommunications, a person or company who has telephone service provided by a telephone company is a *subscriber*.

Subscribers are the users of Digital Subscriber Line services.

Enjoy life in the DSL-powered home

DSL is a fatter data pipe going into your home, but its impact goes way beyond just faster Web surfing. The high-speed DSL connection makes the Internet come alive on your PC in all its interactive, multimedia glory. The always-on part of DSL service means the Internet is always ready and waiting for any online task. Say goodbye to the disconnected feeling between you and the Internet created by slow, dial-up Internet connections.

DSL transforms the PC into an essential household appliance for your entire family, enabling them to become full participants of all the Internet has to offer. The broadband home allows you to use the Internet as a natural extension of your life as you convert a mosaic of everyday activities into a series of online activities, including the following:

- ✔ **Hunting and gathering information**. A high-speed, always-on connection makes surfing the Web for information, getting news, researching, and downloading software second nature.

- ✔ **Transacting household business**. Your DSL connection gives you full-powered access to online shopping, investing, bill paying, banking, and more.

- ✔ **Telecommuting**. DSL combined with VPN brings all the capabilities of your office network to any PC in your home.

- ✔ **Playing**. High-speed, always-on DSL opens up all the multimedia entertainment richness the Internet has to offer. You can play online games, listen to music, make your own music CDs, view video clips, and more.

- ✔ **Communicating**. You can keep in touch with friends, family, and colleagues using an inexpensive video conferencing kit.

If your household has multiple computers, you can connect them to the same DSL line using a growing number of home networking products. Even DSL service offerings that use a single IP address can be shared easily.

A DSL-powered telecommuter works smarter

High-speed, always-on DSL combined with Virtual Private Networking enables you to telecommute to work in the fast lane. VPN technology allows private data to pass securely over the public Internet. The data traveling between two points is encrypted for privacy as it passes over Internet.

As a DSL-powered telecommuter, you can work smarter from home. You have access to all essential services at your office through VPN, including email, databases, files, and intranet resources. DSL also enables you to stay in the loop at the office using sophisticated tools that simply weren't possible with a dial-up connection, such as Web-based collaboration services and video conferencing.

The high speed of DSL creates the same feel of your office network but from the comfort of your home office. Using a DSL connection, you can work more productively at home — and get the bonus of capturing lost time from commuting on congested highways.

The DSL-powered business has the advantage

The power of a digital, high-speed, always-on Internet connection improves productivity and opens up new possibilities for your business. With a DSL Internet connection, your business can

- **Do more in less time.** Faster Web access through DSL means your employees are more productive.
- **Send and receive email instantly.** You've got mail whenever an email message is received at your email box.
- **Get and send files quickly.** High-speed DSL means quick uploading and downloading of files.
- **Connect multiple computers to a single DSL line.** Using a single, high-powered DSL connection allows you link all the computers in your business through a local area network.
- **Support telecommuters without modems and telephone lines.** Your business can add VPN (Virtual Private Networking) capabilities to your DSL connection to allow secure connections between your office and telecommuters, branch offices, or mobile users. With VPN over DSL, you don't need banks of modems and telephone lines at the office.

  ✔ **Collaborate remotely.** With the power of DSL, you can use video confer-
    encing and collaboration software as a practical solution, saving travel
    time and costs.

  ✔ **Manage your business software.** The Internet is the medium of choice
    for getting new software and updates. DSL lets you get the software you
    need from the Internet quickly.

Time for a DSL Reality Check

Although DSL promises big changes for the way the rest of us connect to the
Internet, it's not a yellow brick road. Deploying DSL is a huge undertaking
affected by a variety of technical, financial, and political factors. No honest
coverage of DSL would be complete without alerting you to the elements that
make up the dark side of DSL technology and its deployment.

This section presents an overview of the main trouble spots in DSL today. Keep
in mind, however, that DSL is a fast-moving technology, and some of these
problems may be resolved as it evolves. Details on these problems and how
they can affect you are covered in relevant chapters throughout this book.

Misleading speed claims

The DSL speed touted by most DSL providers and ISPs is the theoretical capability of the DSL connection your premises through the DSLAM at the CO. But
this portion of your DSL Internet connection is just the tip of the iceberg. The
journey your data traffic takes to get to the Internet involves interconnected
backbone networks. After the data leaves your DSL line and goes to the
DSLAM, it's consolidated with other DSL subscriber data and pushed through
larger data pipes. From the DSL provider's (CLEC or ILEC) network, the data is
passed on to the ISP, which in turn passes the data on to the Internet.

Doing the math: DSL service versus dial-up connections

Even moderate use of a single dial-up Internet
connection can easily run more than $50 a
month for business telephone usage charges
plus the cost of ISP service at $20 a month. A
small office with four PCs and four dial-up
Internet connections could incur a cost of a few
hundred dollars a month for just Internet
access. For about the same amount of money,
your business could convert to a 784-Kbps SDSL
connection to enjoy the benefits of a faster,
always-on Internet connection.

In most cases, these backbone networks don't support the theoretical capacity of all the DSL connections being funneled into them. Selling more DSL capacity than can actually be delivered is called *oversubscribing*. The urge to oversubscribe is a compelling economic issue for both DSL providers and ISPs, and they all play the bandwidth game. Oversubscribing ratios between DSL subscribers and the backbone capacity of DSL provider and ISP networks can range from 10-to-1 to more than 50-to-1. As DSL service gets more popular and prices fall, these oversubscribing ratios might get even worse!

As a DSL customer, you never really know what you'll actually get from an ISP or DSL provider. This is a highly deceptive practice; in any other industry, the FTC would probably step in and force DSL providers and ISPs to disclose real-world speeds, in much the way car mileage is defined by the government. The FCC is too busy trying to keep the telecommunications industry competitive to do anything (for now, anyway) to rectify this growing problem.

Oversubscribing is intensifying as the number of DSL subscribers increases. The biggest abuses are going on in DSL service offerings targeted at consumers. New DSL service offering that use PPPoA or PPPoE promise to bring oversubscribing to new heights.

Many DSL providers are planning even more oversubscribing at the DSLAM and their internal backbone networks. They're selling DSL service without building up their backbone networks to deliver the speed.

Subject to change without notice

DSL is a new data communications service that is in a state of constant change. DSL service packages, user-equipment options, and service pricing are all subject to change without notice. Almost everything goes in terms of today's DSL service offerings. Making the right decision about your DSL service takes patience and persistence. Most of the trends are good in terms of lower service costs, better DSL equipment options, and more choices in DSL service configurations. However, DSL — like any new technology — has growing pains.

DSL is not everywhere you want it to be

CLECs and most ILECs are aggressively deploying DSL, but the availability of DSL in your area and specifically to your home or business is still a hit-or-miss proposition.

The most heavily populated areas of most large metropolitan areas have DSL service, often by more than one DSL provider. In many smaller cities, towns, and rural communities, however, the lack of concentration of potential customers discourages ILECs and CLECs from deploying DSL service. In addition, many subdivisions and business parks are also without DSL because their telephone service is commonly provided by DLCs (Digital Loop Carriers) and these devices aren't compatible with DSL. However, technological solutions are available that will be implemented after the costs come down and telephone companies deploy them.

Lengthy installation lead times

DSL installation can be a confusing and complex process for DSL subscribers. The lead times for getting DSL service currently ranges from two to six weeks — and longer in some areas.

Some forms of DSL service from a CLEC take longer than others. SDSL service, for example, requires a new line, so the CLEC must wait for the ILEC to install a new telephone line to the customer premises. The process of ordering the new line from the ILEC and then having the ILEC installer come out to the customer premises can add 10 days to weeks to the installation of the DSL service. After all, ILECs are competitors to CLECs, so they have little incentive to complete installations on time.

After the line is installed by the ILEC, the CLEC sets up the DSL service by sending a technician who does the inside telephone wiring, sets up the DSL CPE, and tests the connection. The entire process can take up to six weeks if no technical glitches are encountered and a lot longer otherwise. The two-tier installation process means you have two installers visiting your premises during the installation process, one from the ILEC and another from the CLEC.

The good news is that most consumer DSL is moving to a self-installation product. Combine this with line sharing, and the long lead times should become dramatically reduced.

Complicated DSL ordering process

DSL service is typically sold as a complete Internet access package, and many ISPs do a poor job of explaining your service options and what they mean to you. Before you order your DSL service, you need to make several decisions to tailor the service to your needs. Here is just a sampling of the kinds of questions you need to answer to determine what service is right for you:

✔ **What type of DSL service do you want to use?** Different DSL flavors have different capabilities, and different DSL providers offer different speed configurations. Typically, you'll be deciding whether to use ADSL or SDSL.

✔ **What speed do you need or want?** A number of factors come into play when deciding how much speed you need for your Internet connection, such as the number of people sharing the service and which applications you plan to use with the service.

✔ **Do you want bridged or routed service?** DSL modems (bridges) and routers are internetworking devices used to connect your PC or LAN to the Internet through your ISP. Each device has advantages and disadvantages.

✔ **What IP configuration do you want?** The Internet Protocol (IP) forms the basis of the Internet and is at the heart of configuring your PC or LAN to connect to the Internet. The core elements of IP networking are IP addresses and DNS (Domain Name System).

✔ **Who will host your email service?** Email is an essential component of any Internet connection. The ISP providing your DSL service usually hosts your email.

Keeping your distance

DSL is distance sensitive due to the higher frequencies DSL uses to enable higher data speeds. Depending on the DSL flavor, distance limitations typically run from 12,000 to 18,000 feet between the customer's premises and the central office. This distance is based on the snaking path that the copper link takes, not as the crow flies.

By most estimates, 80 percent of the United States population lives close enough to a central office to take advantage of the more popular DSL technologies. DSL technologies continue to improve, so today's distance limitations are by no means hard and fast.

True competition is a journey

Even with the legislative push of the Telecommunications Act, ILECs continue to wield tremendous power in shaping DSL service and deployment in their respective service areas. They have the legal and lobbying resources to challenge the government and the implementation of all the provisions of the Telecommunications Act.

ILECs don't make it easy for CLECs for deploy DSL service. The process for CLECs to gain access to an ILEC's CO space and local loops is a long and arduous journey. From the DSL consumer perspective, the success of CLEC competition is critical to ensuring competitive prices and service options. Compared to ILECs, data CLECs are generally smaller, faster moving, technology-driven companies that use their speed and capabilities to deploy innovative data communications solutions.

It's a tedious process to move towards a truly competitive marketplace, but great strides have been made. The FCC and many state regulatory agencies are committed to seeing competition flourish in telecommunications. A good example of the continuing evolution of competition in DSL deployment is the recent line-sharing ruling by the FCC, which allows CLECs to offer DSL to the residential market to compete with the ILEC. By allowing CLECs to offer DSL service over an existing line, consumers will have real competition in the delivery of DSL to their homes.

Waiting for standards implementation

A pivotal milestone in the deployment of affordable, off-the-shelf DSL equipment (referred to as CPE, or Customer Premises Equipment) is the emergence of standards. DSL equipment interoperability is an essential factor in lowering the cost of CPE and increasing consumer choice. With multiple vendors using the same standards for DSL service, consumers can choose the best CPE for their needs. (A *protocol* is a set of rules for data communications. A *standard* is a set of detailed technical guidelines used to establish uniformity. Protocols and standards create an environment of universal capability.)

A lot of progress is being made in developing standards for the leading DSL flavors, but none have been deployed as of this writing. However, these standards should form the basis of DSL deployments in late 2000. The two ADSL standards are G.dmt and G.lite.

The G.dmt standard supports up to 8 Mbps downstream and 800 Kbps upstream. G.dmt is actually the nickname for the standard officially known as ITU-T Recommendation G.992.1. The G.lite standard can deliver 1.5 Mbps downstream and 512 Kbps upstream. G.lite is a nickname for the standard officially known as ITU-T Recommendation G.992.2.

A new emerging standard for SDSL, called G.shdsl, is a rate-adaptive variant of HDSL-2 and promises to become the future replacement for the current SDSL.

What's on Cable?

No discussion of DSL service is complete without covering its main competitor, cable modem service. The second largest network in the United States is cable television, which passes 90 percent of all homes. The cable companies sit on a huge source of bandwidth for Internet connections. The cable industry has been aggressively deploying high-speed cable modem access and currently leads DSL in both homes passed and number of subscribers. By the end of 1999, 35 to 40 million homes had been passed by cable modem service, with 1.7 million subscribers.

Cable service is delivered as a complete package sold through your cable operator. Cable Internet service uses a cable modem that connects to a different coaxial cable inside your home than the one used for for your cable TV service.

Cable modem service is delivered using a different architecture than DSL, and this difference can affect the performance of cable. Cable service is a shared connectivity medium much like a local area network. The cable system architecture connects hundreds or thousands of homes into large Ethernet-like networks. As more users are connected to the cable network in a given area, the throughput of the connection slows down for everyone. The often-hyped rates of 1.5 Mbps or higher downstream and 300 Kbps upstream quickly evaporate in this shared architecture.

Should you check out cable modem service?

Cable modem service is primarily a consumer — rather than a business — Internet access service because most residential areas are wired for cable service and most business locations are not. Most cable modem service providers don't provide static IP addresses or support DNS services. Most cable modems are designed to connect only a single computer to the service. You can get around this limitation, however, by sharing the cable connection across a LAN using a proxy server or an Ethernet router.

If your Internet access and bandwidth needs fit the constraints of cable modem service and DSL service isn't available in your area, you might want to consider cable service (if it's available in your area). A typical cable modem service costs $30 to $40 a month, which includes the cable connection and Internet service. You also typically pay a one-time installation charge of about $100, and you may be charged for the cable modem.

Cable competition can't hurt

Cable operators offering cable modem service represent a major competitive force to the consumer DSL offerings of the telephone companies. Cable companies, with their aggressive pricing and early leadership in taking a new approach to installing a high-bandwidth service, have forced the telephone companies to do things differently.

Cable companies were the first to buck conventional wisdom in terms of going to the customer's site and getting the entire Internet connection up and running. They eliminated the segregation between computing and data communications, which has been the way telephone companies have treated their data communications service.

The cable industry had a year's head start in deploying affordable, high-speed Internet access. Cable companies have delivered their cable modem service in a user-friendly way that challenges the telecommunications industry to do the same. When a user orders cable modem service, they come to your home, connect the cable modem to your PC through an Ethernet connection, and even install a network interface card if you don't have one. They do all the cabling. When they leave, you're up and running. Cable modem service is here to stay, and if it fits your needs, go for it.

Chapter 2

The Telecom Side of DSL

DSL is the data communications conduit between you and your Internet service provider. Although the DSL part of your connection is transparent, its operation plays a big role in defining your DSL Internet service options. This chapter takes you on a guided tour of the telecommunications side of your DSL connection to understand how it affects you.

Welcome to the Telecommunications Jungle

The telecommunications industry is in the throes of a massive upheaval. A convergence of legislative, technological, and competitive forces is fueling changes in the way telecommunications services are delivered. With change comes large-scale confusion for telecommunications providers, Internet service providers, and consumers. In the long run, however, the unleashing of competitive forces into the telecommunications industry is in the consumer's best interest.

DSL service deployment is heavily influenced by these changes. Not only is DSL a new data communication technology, it's also spearheading the new era of competition in the telecommunications industry. Understanding how the telecommunications industry operates is an essential starting point in your journey toward DSL enlightenment.

Gatekeepers of the last mile

AT&T, which was known as the Bell System, was once the only telephone company for most of the United States. In 1984, the United States government forced AT&T to divest itself of its local telephone companies. The local operations of AT&T were broken up into Regional Bell Operating Companies (RBOCs). One of the most significant changes mandated by the breakup of AT&T was the redefinition of the nation's telecommunications system into local exchange carriers (LECs) and interexchange carriers (IXCs).

An LEC is the telephone company that provides local telephone service. An IXC is a long-distance telephone company. This forms the basis of the telephone system today. Every home and business in the United States with telephone service goes through an LEC. Long-distance calls are routed from the LEC to the long-distance telephone company and then to the LEC at the destination.

LECs include both RBOCs (Regional Bell Operating Companies) and independent telephone companies (ITCs). There are close to 1400 independent telephone companies. The dominant players in local telephone service are the telephone companies that provide your voice service. You know these telecommunication companies as Ameritech (SBC), Bell Atlantic, BellSouth, GTE, Pacific Bell (SBC), Southwestern Bell (SBC), and US WEST.

LECs provide the connections to almost every telephone user in the United States. They control the copper wiring that connects almost every home and business within a given geographical area to an LEC facility called a central office (CO). This wiring, called a *local loop,* consists of the twisted-pair copper wiring used for POTS (Plain Old Telephone Service) connections between central offices (COs) and customer sites. DSL service comes into play where these local loops connect to central offices. Local loops — the lines between an ILEC's CO and the telephone subscriber's premises — are commonly referred to as the *last mile* of the telecommunications network, although the distance of a local loop can stretch out to more than three miles from the CO. The control of the last mile by LECs operating as monopolies has been the largest single factor in the high cost of bandwidth for data communications.

Once there were seven; now there are three

The Telecommunications Act removed many barriers against the merger of RBOCs (Regional Bell Operating Companies). In an ironic twist, the 1984 break up of AT&T into regional telephone companies has been reversed as RBOCs have merged into larger entities. Southwestern Bell Corporation (SBC) now owns Pacific Bell and Ameritech. Bell Atlantic has merged with GTE, and US West has been acquired by Quest Communications. Only Bell South remains as an independent RBOC.

Changing the rules

The ultimate goal of the Telecommunications Act of 1996 is to open up local telecommunications services to competition. In the parlance of the Telecommunications Act, LECs became known as *Incumbent Local Exchange Carriers (ILECs)* because they control the local telephone service infrastructure (COs and local loops). The new telecommunications players are called *Competitive Local Exchange Carriers (CLECs)*.

At the heart of DSL deployment by CLECs are the Telecommunications Act's provisions covering unbundled access and co-location. These network elements include central office space, transmission between COs in a metropolitan area, and wiring systems used to deliver telecommunications services to customers. Using the interconnection provisions of the Telecommunications Act, CLECs can install their DSL equipment (called Digital Subscriber Line Access Multiplexers, or DSLAMs) at the COs owned by ILECs. CLECs then lease local loops that terminate at the CO to provide DSL service to customers.

Through the regulatory maze

The Federal Communications Commission (FCC) along with 50 state regulators define the rules for the delivery of telecommunications services. Telecommunications oversight, policy, and regulation at the federal level is the province of the FCC. These functions at the state level are performed by each state's regulatory agencies, typically called public utility commissions (PUCs). These state agencies add another layer of regulations for telecommunications at the state government level and exercise regulatory powers over ILECs and CLECs. As a check on state regulatory agencies, the Telecommunications Act disallows any state agency from prohibiting any qualified entity from providing interstate or intrastate telecommunications service.

The Telecommunications Act has created ongoing confusion on jurisdiction between the FCC and state public utility commissions on the control of in-state pricing of ILECs and CLECs. The details of the implementation of the Telecommunications Act and the enforcement of its provisions were left, for the most part, to the FCC. The FCC has made a number of rulings related to the implementation of the Telecommunications Act, and challenges to many of them are tied up in the courts. The main message here is that the Telecommunications Act is an evolving framework subject to changes, as challenges in courts become formalized.

The Telecommunications Act has taken root in the telecommunications industry and especially in the delivery of DSL. The United States Congress could change this law, however, to favor ILECs. The FCC, as the primary government agency responsible for carrying out the Telecommunications Act mandate, continues to remove barriers to competition in the delivery of DSL but it continually faces threats from a US Congress that receives large campaign funding from the ILECs to strip its authority.

In November 1999, the FCC ordered ILECs to allow line sharing. This ruling promises to level the playing field for the delivery of DSL to consumers. Before this line-sharing ruling, ILECs didn't allow CLECs to offer DSL service over existing voice lines. As a result, businesses and consumers had to either use the ILEC DSL offerings or bring a new line in for DSL from the CLEC, which made the service considerably more expensive.

Inside the Telecommunications Network

DSL service is a data communications service that operates in the realm of the telecommunications network of central offices and local loops but uses them differently than the way they're used for voice communications. The COs and local loops become the gateway to a data communications network and the voice communications network. The PSTN (Public Switched Telephone Network) is a wide area network on a global scale. Just about every business and home connects to this network through twisted-pair copper wires, which carry both voice and data communications.

All loops lead to the central office

The central office (CO) is the front line of the telecommunications network, as shown in Figure 2-1. Think of a CO as a network node on the PSTN. All local loops (telephone lines) from every home and business terminate at a CO that services a specific geographical area. The term *local loop* is a generic term for the connection between a customer's premises and the CO.

The United States has more than 19,000 COs, which terminate more than 200 million local loops. (An estimated 700 million local loops exist worldwide.) Large metropolitan areas typically have a number of COs, which can be in an entire building or a room in a city building. These central offices connect to other COs for local calling or to other switching facilities for long-distance calls.

The common symbol used to represent any wide area network is a cloud, as you can see in Figure 2-1. The concept of a cloud relates to the fact that the operation of the network is hidden from the user.

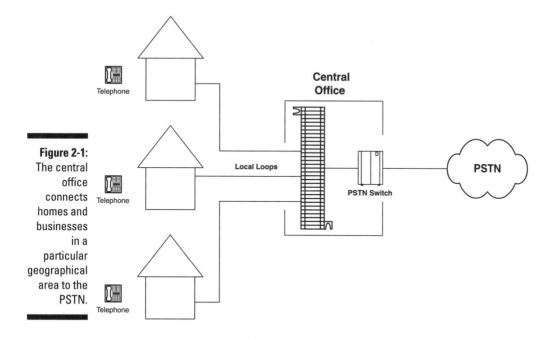

Figure 2-1:
The central
office
connects
homes and
businesses
in a
particular
geographical
area to the
PSTN.

Subscribers are tethered through local loops

If you have a telephone, you're connected by a local loop between your premises and the CO. The local loop connects your premises to the PSTN through the CO. Local loops are also called *access lines* because they enable users to access the CO. The majority of local loops are within 18,000 feet of a CO.

At the CO, local loops end up in the Main Distribution Frame (MDF), which is the wiring center for local loops. For POTS services, the wiring from the MDF is connected to the DSLAM and PSTN switch. This is accomplished through the use of a splitter, which separates voice and data to send them to their respective network.

From the PSTN switch, wiring is carried on telephone poles or buried in the ground until it reaches a local interconnection point near the customer's premises. This interconnection point is a box on a telephone pole in a neighborhood or a wiring closet in an office building.

From this interconnection point, the wires are routed to individual sites and end up at a network interface device (NID). In most homes, a NID is a small box located inside or outside the home at the point where the telephone line(s) enter the home. The NID is the demarcation point between the telephone company side of the network and the inside wiring for your home or office.

Extending the reach of local loops

Digital Loop Carriers (DLCs) extend the reach of telecommunications services from a CO. DLCs are typically used in office parks and housing developments to minimize the need to run local loops over several miles to the CO servicing the area. Instead, local loops connect a cluster of homes or businesses to a remote terminal (RT), which in turn concentrates the traffic into a higher bandwidth for delivery to the CO, as shown in Figure 2-2. Remote terminals are typically green boxes located on a telephone pole or larger refrigerator-sized boxes on the ground.

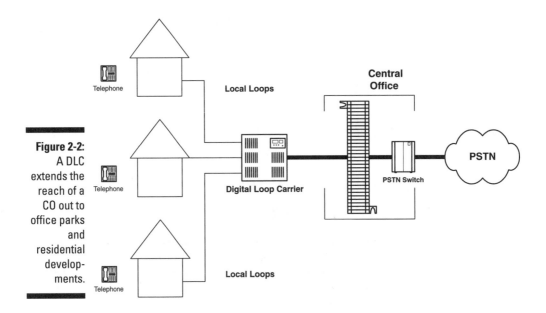

Figure 2-2: A DLC extends the reach of a CO out to office parks and residential developments.

By terminating loops at the DLC remote terminal, you reduce the effective length of the local loop and thus improve the reliability of the service. This architecture solved many problems for POTS, but it causes problems for the delivery of DSL services. The number one problem with DLCs is that the back (network) end is fiber, and DSL cannot travel over fiber. So the DSLAM must be in the DLC, and DLCs were not designed for this.

Cards that offer DSLAM-like functionality and can slide into the DLCs must be made available. This has taken time, and only now are solutions to this problem being deployed. This is a critical availability issue because in some ILEC territories, DLCs carry almost 50 percent of the traffic on average. DLCs cover 25 percent or so of the local loops in the United States. The DLC last-mile problem will fade away after the ILECs make the commitment to spend the resources to solve it with new equipment.

The big PSTN switch

The Public Switched Telephone Network (PSTN) uses digital switches (often called a Class 5 switch) to route telephone calls, with telephone numbers acting as a routing address system. *Switched services* are those in which the connection is made by some call control procedure, such as a person dialing a call to another telephone.

The PSTN uses circuit switching to transmit calls. A *circuit* is a physical path for the transmission of voice and data. The communications pathway remains fixed for the duration of the call and is not available to other users. A *circuit-switched connection* between two users becomes a fixed pathway through the network. At the end of the communications session, the pathway is terminated and is available to other users. In telecommunications lingo, the DSL service itself is often referred to as a *circuit.* The DSL service is a point-to-point circuit. This type of service is referred to as *dedicated, always-on, or nailed up.*

The Internet, the PSTN, and DSL

PSTN switches at the COs are sized according to probability. If everyone in the United States picked up their phone at the same time, only a small portion would be able to make calls. These switches are designed with the idea that a subscriber uses his or her phone line only a few times during the day. Therefore, instead of a 1-to-1 relationship between the customer's telephone line and the network, it's more like a 1-to-4 or 1-to-6 relationship.

The growth of the Internet has wreaked havoc on this probability environment because people are dialing up to their Internet provider and staying online for a long time — a lot longer than ever envisioned. As a result, telephone subscribers in high-Internet subscriber areas get poorer basic telephone service. For example, a switch designed to support 10,000 lines coming from customers' homes might support only 400 outbound lines from the CO for these customers to use. If 100 Internet subscribers stay online to surf the Internet, the availability of lines for everyone else is reduced by 25 percent — a large amount in the telephone industry.

This offloading of data traffic from PSTN switches to DSL lines connected to DSLAMs frees up PSTN switches for voice traffic. For ILECs, this is one of the big advantages of DSL.

Inside the DSL Network

DSL data communications use key elements of the telecommunications network, most notably local loops and central offices. For ADSL and G.lite connections with both data and POTS service on the same line, voice and data are separated. The POTS traffic goes to the PSTN switch to be handled as any telephone call. The data part goes to the DSLAM. The following sections explain the elements that make up the DSL system.

What's a DSL provider?

The *DSL provider* is the company delivering the DSL service to your home or business through the telephone line. This can be a CLEC, an ILEC, or an ISP acting as a CLEC. DSL providers are the companies that build the DSL network within a regional market or on a national scale. Competition at this level involves competing DSL networks, where DSLAMs are installed on a CO-by-CO basis and then linked into a network operated by the DSL provider.

The DSL network build out is driving DSL deployment. Because of the competition among DSL providers (CLECs and ILECs), they have built their networks in the same areas. Most large metropolitan areas have more than one DSL provider; many have up to four companies offering DSL service. Diversity of DSL providers provides a vibrant environment for choices, prices, and innovation.

DSL networks operated by CLECs and ILECs are the power grid for the next-generation Internet. These high-speed networks represent the new infrastructure for supporting a new way of doing things on the Internet for businesses and consumers.

The customer interface for these DSL providers is the Internet service provider. Many DSL providers also act as their own ISP. Most ILECs operate their own ISP organizations that provide the customer interface to their DSL service. CLECs typically use ISP partners to sell their DSL service although some also sell direct.

DSLAMs are the center of the DSL universe

COs house *DSLAMs* (Digital Subscriber Line Access Multiplexers), which consolidate data traffic from individual DSL connections into large high-capacity backbone networks. Figure 2-3 shows the CopperEdge 200 concentrator, a DSLAM from Copper Mountain Networks. The DSLAM is a platform for DSL service aggregation.

When a CLEC or an ILEC sets up DSLAM devices at a given CO, they make DSL service available to the area serviced by that CO. Not everyone in the area may actually be able to get the DSL service because of technical limitations.

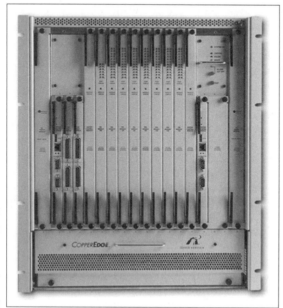

Figure 2-3:
The CopperEdge 200 DSL concentrator.

Photo courtesy Copper Mountain Networks, Inc.

DSLAM vendors include Alcatel, Copper Mountain, Nokia, PairGain, Cisco, and others. Copper Mountain supplies DSLAMs to NorthPoint Communications and other CLECs. Nokia supplies DSLAMs to Covad and others. Rhythms uses Paradyne, Copper Mountain, and Cisco. Alcatel supplies DSLAMs to most ILECs.

A CO is not the only place for DSL capability

Multiple Tenant Units (MTUs) is telecommunications jargon for a high-rise apartment or office building. Many of these larger buildings are signing deals to bring high-speed data communication service into the building and put a DSLAM in the basement to carry high bandwidth up to the tenants in the building.

Is your loop qualified?

Loop qualification is the process used to determine whether a specific copper pair will support DSL. To start the process, you provide your 10-digit telephone number for the location where you want DSL service. (COs have their own addresses — the first six digits of your telephone number, which are the area code and exchange office code.) Typically, this process verifies whether the central office has a DSLAM installed. Adding the specific address of the premises where you want to install DSL service defines the approximate loop length between your premises and the CO.

ILECs do automated loop testing in most of their COs, which allows them to determine fairly accurately whether DSL will work and use actual measurements, not distance "as the crow flies," to tell you whether your line qualifies. But if your premises is within 18,000 feet of a CO and you don't qualify for DSL service, request a manual line test, which is more accurate. If the problem on the line is a load coil or a bridge tap, it can usually be cleared or a separate line can be used — if the telco can be persuaded to cooperate.

A variety of factors go into determining whether a local loop can support DSL, including the following:

- ✔ **DSLAM used at the CO.** Different DSLAMs supporting different DSL flavors have different capabilities.

- ✔ **Local loop wires.** Many are 24 or 26 AWG (American Wire Gauge). The AWG measures the thickness of the copper wiring. The thicker the wire, the less resistance it has for signals traveling over it. The thicker the copper wiring, the longer the distance that DSL service can be delivered.

- ✔ **Whether loading coils have been placed on the loop to improve voice quality on longer loops.** A loading coil is a metallic, doughnut-shaped device used to extend the reach of a local loop beyond 18,000 feet for POTS. Unfortunately, loading coils wreak havoc on DSL. If the local loop has any loading coils, they must be removed to use DSL.

- ✔ **Whether a bridge tap has been added to the local loop.** A bridge tap is an extension to a local loop generally used to attach a remote user to a central office without having to run a new pair of wires all the way back. Bridge taps are fine for POTS but severely limit the speed of DSL service. They can often be found and fixed.

- ✔ **Spectrum incompatibility in a binder bundle.** The packaging of many copper wire pairs into a binder bundle has implications for the delivery of DSL service due to spectral interference (also called *crosstalk*), which happens when neighboring lines are corrupting each other.

Going the distance

Because of the physics of high-speed data communications, DSL service is distance sensitive. The wire distance between your premises and the CO plays a critical role in determining whether DSL service can be delivered to you and at what speeds it can be delivered. The wire distance is often different that the actual distance of your premises from the CO.

Most DSL technologies have a distance limitation of within 18,000 feet (3.4 miles, or 5.5 km). By most estimates, 60 to 70 percent of the population of the United States live close enough to a central office to take advantage of the more popular DSL technologies. The largest segment of the population that can't get DSL is supported by Digital Loop Carriers (DLCs).

The following table shows the maximum wire distance supported by the different DSL flavors. Remember, though, that distance is not the only factor that affects whether DSL service is available:

DSL flavor	Distance limit
ADSL	18,000 feet
G.lite	18,000 feet
SDSL	18,000 feet
IDSL	36,000 feet
VDSL	4,000 feet
HDSL	15,000 feet

End of the Line

This section presents an overview of the DSL equipment and inside wiring that make up your end of the DSL connection.

Customer Premises Equipment (CPE)

In telecommunications lingo, the generic term for DSL hardware used at your premises is *Customer Premises Equipment (CPE)*. CPE devices provide two distinct functions for the DSL and Internet connection. The first is the Channel Service Unit/Data Service Unit (CSU/DSU), which provides the termination for the DSL digital signal and handles the transmission of the signals over the DSL line. The second is to handle the Internet Protocol (IP) networking functions of the Internet itself, which passes over the DSL line.

The two types of DSL CPE are DSL modems, which are bridges, and routers. These internetworking devices allow separate networks to communicate with each other. In the case of DSL Internet service, a modem or a router acts as your gateway to the ISP's network and on to the Internet.

A *DSL modem* seamlessly combines two networks into a single network. Bridges are less expensive and simpler than routers because they require no special configuration by the customer. Essentially, using a bridge makes your computer or LAN part of the ISP's TCP/IP network. However, bridges lack the enhanced networking and security capabilities of a router. Bridges are typically used for consumer and SOHO (small office/home office) offerings.

A *router* forwards data traffic between separate networks. Routed service means your local network is defined as a separate TCP/IP network from the ISP's network. A router examines network addresses in the data packets it receives and forwards to the ISP's router data destined for the Internet. A router brings to your Internet connection enhanced internetworking features, including NAT (Network Address Translation), DHCP (Dynamic Host Configuration Protocol), and security features. Figure 2-4 shows the DSL router from Netopia.

DSL CPE can connect to a computer using Ethernet, USB (Universal Serial Bus), or as a PCI adapter. See Chapter 8 for more details on DSL CPE options.

Figure 2-4: The Netopia router.

Photo courtesy Netopia, Inc.

The great divide

In the United States, telephone wiring responsibility is divided between the telephone company, which is responsible for local loop wiring to your doorstep, and you, the customer, who is responsible for any inside wiring. In homes, a *network interface device (NID)* usually separates the telephone company's wiring and your premises wiring.

Figure 2-5 shows a network interface device. The NID allows the telephone company installer to bring the line to your home. From this box, you or the telephone company installer attaches the twisted-pair wiring that leads to the specific location of your CPE.

Figure 2-5:
A Siecor
network
interface
device
(NID).

Photo courtesy Siecor

If your premises are in an older structure, you might have a *protector block* instead of an NID. The FCC prohibits customers from working at the protector block, so the telephone company must install an NID for you to get DSL service. The FCC also has the *12-inch rule,* which states that if you have a protector block, the NID must be located within 12 inches of it.

Getting wired for DSL

POTS and DSL service require the same physical wiring. Depending on the DSL flavor installed, you might need to get a new line installed or you might be able to add DSL service to an existing voice telephone line.

Standard telephone wiring consists of a single 4-wire cable, which allows for two lines. For residences outside metropolitan areas, the number of telephone lines might be restricted to two.

For your inside wiring from the NID to the location where you plan to use your DSL CPE, the installer of your DSL line (ILEC or CLEC) might do the inside wiring. DSL service is handled differently than traditional wiring because the installation of the service typically involves a certain specified amount of additional inside wiring to bring the line to the DSL CPE location. To what extent inside wiring is handled by a DSL provider varies.

In most earlier ADSL deployments, a DSL provider installed a *POTS splitter,* which separates the POTS channel from the high-speed DSL channel. Most ADSL and G.lite installations today are splitterless, which means no POTS splitter is installed at your premises. However, G.lite might require the installation of a small filter device, shown in Figure 2-6, next to every POTS device (telephone, fax, or modem) sharing the G.lite line.

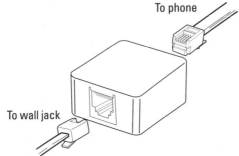

Figure 2-6:
The EZ-DSL
microfilter.

Know your connectors

At the end of your wiring are connectors. The connector jack used for your DSL wiring can be either an RJ-11 or an RJ-45 modular connector, which snaps into a jack. You can plug an RJ-11 connector into an RJ-45 jack. Figure 2-7 shows the RJ-11 and RJ-45 connectors. Most residential service uses standard RJ-11 jack, and many businesses with PBX-style telephone networks use RJ-45 jacks.

Figure 2-7:
The 4-wire
RJ-11 and 8-
wire RJ-45
modular
connectors.

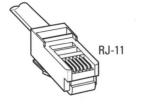
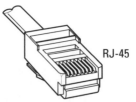

DSL Service from CLECs

DSL CLECs are commonly referred to as *data CLECs*. Whatever they call themselves, these CLECs are conversant in the language of the local loop but also grounded in the world of the Internet and IP networking. The leading national DSL data CLECs are Covad Communications, NorthPoint Communications, and Rhythms NetConnections. In addition, there are a growing number of regional CLECs.

CLECs are leading the charge in the delivery of DSL services in many areas. By the end of 1999, most large metropolitan areas were served by competing CLEC DSL networks. For 2000, the leading CLECs have big plans to build out their networks to thirty second-tier metropolitan areas.

The leading data CLECs have focused on delivering business-class data communications services through SDSL (Symmetric DSL). With the FCC line-sharing ruling, CLECs will be aggressively launching ADSL and G.lite to the home market, competing head on with ILECs. Before this ruling, ILECs would not allow CLECs to provide DSL over their existing voice lines.

How a data CLEC operates

Within each CO, the CLEC maintains DSLAMs in racks that connect to unbundled local loops supplied by the ILEC. The CLEC works with the ILEC to get the twisted-pair wiring delivered to the network interface device (NID) at the customer's premises. After the ILEC installs the twisted-pair, the CLEC installer does the inside wiring, tests the line, and sets up the DSL CPE.

The CLEC maintains a backbone network center that acts as the central hub for data traffic coming from DSLAMs in multiple central offices throughout a metropolitan area. The ISP taps into the CLEC's central hub through a single fat pipe to move IP traffic.

The CLEC defines which areas have DSL service available by determining which COs get DSLAMs. The CLEC manages all CPE installations, maintains the local loop and co-location sites, and monitors and manages the internetwork connecting all COs to the central hub.

CLECs, ISPs, and DSL service

The relationship between most CLECs offering DSL service and the ISP is a partnership. The CLEC makes the DSL service available to the ISP on a CO-by-CO basis. The ISP typically buys the circuit for a fixed cost based on the speed of the connection. The ISP provides the customer interface to the combined DSL data communications link and Internet access service. The ISP offers the IP networking services, including Domain Name Server (DNS) services, mail services, Web hosting, and any other customization services.

Most of the big data CLECs (Covad, NorthPoint, and Rhythms) are building their own identity in the DSL market so that DSL customers will ask for their particular brand of DSL network service.

The ISP ultimately creates the actual DSL service package and pricing. The cost of the CLEC's DSL service, commonly called the *DSL circuit,* is typically included in a single bill from the ISP. Although CLECs typically install the link at the customer's premises, the ISP configures the DSL CPE, which is shipped to the premises before the DSL service is installed.

The downside of this partnering relationship is that the consumer gets uneven DSL service offerings and service depending on the ISP. Many ISPs do an excellent job of marketing and customer service for DSL, but far too many do a poor job. As a customer of DSL service, you need to compare different ISP service offerings even though the DSL service might come from the same CLEC.

DSL Service from ILECs

Most ILECs are deploying ADSL because it supports high-speed data communications and POTS service on the same line. Simply put, with ADSL you can have high-speed Internet access and regular telephone service on one line. ILEC DSL service offerings are typically broken down by speed, with the lower speed offerings targeted at the home market and the higher speed offerings targeted at the business market.

ILECs are offering DSL service using two methods. The first method is one in which the ILEC partners with an ISP, who then resells the ILEC's services. This means you can get ILEC-resold DSL service through ISPs. The DSL portion of the Internet access package is typically (but not always) added to the single bill from the ISP. The second method for offering DSL service is as a bundled service with Internet access, provided by both the ILEC and its in-house ISP, such as PacBell.net.

If you get a new line, you'll be required to sign up for POTS service as well, because most ILEC DSL implementations are ADSL that supports POTS. This means you'll be paying additional line charges beyond just the DSL service charge.

Chapter 3

The Zen of DSL

*B*y understanding the basics of bandwidth and data communications, you build the foundation you need to understand the lingo and navigate through the maze of DSL technologies. This chapter provides an overview of the DSL technology that makes it possible to use copper telephone lines to deliver high-speed data communications.

DSL Data Communications 101

DSL service takes you into the world of digital data communications. Terms such as *digital, bandwidth, asymmetric,* and *symmetric* are spoken in the everyday world of DSL. This section explains the essential concepts and terminology associated with DSL data communications. Grasping these buzz words will allow you to speak the local language to better navigate the world of DSL service.

The nature of digital communications

Digital communication is the exchange of information in binary form. Unlike an analog signal, a digital signal doesn't use continuous waves to transmit information. Instead, a digital signal transmits information using two discrete signals, or on and off states of electrical current. All computer data can be communicated through patterns of these electrical pulses. Digital communication can support high-speed data communications and is more reliable than analog communication. Because DSL is a digital form of data communications, your Internet connection is more reliable.

Digital data may be represented in many ways, so a special interface device must convert digital content from the computer to a form of digital signaling used for the data communications link. The conversion of data is the reason many DSL CPE products are called DSL modems. The term *DSL modem,* however, is more a marketing term than a technical one. DSL modems do not perform the modulation/demodulation performed in analog modems but they do have a modem-like functionality in making the connection to the Internet through the DSL WAN (wide area network).

Higher frequencies mean more bandwidth

The capacity of a data communications line is based on the medium and the frequency range used by a given data transmission technology. *Media* is the stuff on which voice and data transmissions are carried, which in the case of DSL are the twisted-pair copper wires that make up the local loop. *Bandwidth* refers to capacity as defined by the combination of the media used and the frequency spectrum that the media can support. *Frequency* equals the number of complete cycles of electrical current occurring in one second, which is measured in Hertz, or cycles per second. The wider the frequency used, or Hertz, the greater the capacity, or bandwidth.

The use of higher frequencies in DSL to support higher data communications results in shorter local loop reach. This is because high-frequency signals transmitted over copper loops dissipate energy faster than lower frequency signals. The electrical properties of copper wiring create resistance and interference problems with data transmissions. This forms the basis of the inherent limitations of DSL service based on the distance between a customer's premises and the central office servicing a given area.

The concept of frequency is important in understanding how the same telephone line can be used for analog voice service and yet also support higher bandwidth connections at the same time. For example, ADSL can use the same line as POTS because ADSL exploits a higher frequency than that used for voice service. POTS carries voice communications in the voice frequency range of 300 to 3.3 kHz. ADSL uses a much broader range of frequencies that range from 32 kHz to 1.1 MHz.

Getting your signals straight

Signaling is the process whereby an electrical signal is transmitted over a medium for communications. To transmit data across the twisted-pair copper wiring between your premises and the CO, some method of signaling must be used. High-frequency signals transmitted over copper loops dissipate energy faster than lower frequency signals. The loss of higher frequencies is one reason potential bandwidth goes down with distance. Modulation techniques minimize the loss of electrical energy as it passes over a copper wire by reducing the frequency, which in turn extends the local loop reach. Different DSL flavors use different modulation techniques. For example, IDSL uses a modulation called 2B1Q. ADSL uses a modulation called DMT (discrete multi-tone).

Going asymmetric or symmetric

In many forms of data communications (including DSL), one data channel may support a larger data communications capacity than the other channel. This unequal form of data communications is referred to as *asymmetric,* which means data traveling in one direction moves faster because of a higher capacity than data moving in the opposite direction. In the case of Internet access, an asymmetric connection means data coming from the Internet to your computer travels at a much higher rate than data going from your computer to the Internet.

In *symmetric* data communications, data moves at equal speeds in both directions. The DSL family of technologies has both symmetric and asymmetric versions.

Note, though, that asymmetric versions of DSL can support symmetric services by matching the downstream speed with the maximum upstream capability of the DSL service.

Be careful about some ADSL/G.lite service offerings that severely restrict the speed of data going from your computer to the Internet. One ILEC has an ADSL offering that supports only 90 Kbps upstream.

Data swimming upstream or downstream

The carrying capacity of a communications link is measured by the two directions that data moves: upstream and downstream. *Downstream* refers to data traffic moving from the Internet or another remote network to your computer or network. *Upstream* refers to data moving from your computer or network to the Internet or another remote network.

If you upload a file to an FTP server on the Internet, for example, the upstream capacity of your DSL connection determines how fast the file will be uploaded. Likewise, if you download a file from an FTP server, the downstream capacity of your connection determines how fast the file is transferred. For DSL service, you see references to downstream and upstream speeds as, for example, 1.5 Mbps/384 Kbps, which means the service offers 1.5 Mbps downstream and 384 Kbps upstream.

Throughput (Bandwidth in the real world)

Bandits of bandwidth lurk on any data communications link. These bandits come from a variety of factors that affect the true throughput of a connection. *Throughput* is an overall measure of a communication link's performance in terms of its real-world speed.

Having the fastest link to the Internet is only half the story of what your real throughput will be. Remember, the Internet itself is one giant, shared network. It suffers from rush hour traffic just as highways do. With heavy network traffic and a crowded Web server, your speed can dramatically slow down.

A number of factors can affect your connection and make the throughput figure lower:

- ✔ **Internet traffic volume between your computer or network and the remote server.** If the Internet is experiencing a heavy volume of data traffic, the entire system slows down.

- ✔ **The speed of the server handling your request.** If the server is busy, it will slow down in terms of delivering your requested data. With faster DSL connections, you begin to notice the slower Web servers, something you might have never noticed when you used an analog modem.

- ✔ **The quality of your DSL connection to the ISP.** A noisy telephone line can require data speeds to be lowered on the fly.

- ✔ **The backbone used by the ISP to connect to the Internet.** If an ISP is oversubscribing DSL customers by not providing enough data communications capacity to the Internet, your throughput will deteriorate.

- ✔ **Oversubscribed DSLAMs at the CO.** Depending on the DSLAM architecture, data traffic may be concentrated into a smaller backbone than the cumulative total capacity of all the DSL lines going into the DSLAM.

- ✔ **Bottlenecks with the DSL provider's metropolitan network.** The DSL provider uses a backbone network that links all the DSL-enabled COs to a central location for pass-off to ISPs. The metropolitan network can have bottlenecks from the COs to the network center.

DSL Has More Flavors Than Ben & Jerry's

Having a variety of flavors when ordering an ice cream cone is great, but when it comes to DSL it can get confusing. Life would be easier if just one form of DSL was available. But who said life was easy?

Each different DSL flavor has different capabilities and configurations. In addition, different DSL providers offer different DSL flavors. Understanding the attributes of each DSL flavor will go a long way in helping you analyze your DSL options.

ADSL

Asymmetric Digital Subscriber Line (ADSL) delivers a range of data communications speeds. Downstream speeds can reach up to 8 Mbps, and upstream speeds can reach up to 1 Mbps. ADSL can also be packaged as a symmetric service.

Because ADSL shares the same line used for POTS service, you can supplement an existing POTS line with ADSL service, which makes it an attractive option for the residential and small business market. The voice and data are split using a POTS splitter. POTS splitters are filters used at each end of the local loop to split the data traffic between low-frequency voice communications and high-frequency data communications. The splitter allows existing analog voice and data services to coexist on the same line as the one used for the high-speed data service.

The G.dmt standard promises to become a basis for ADSL deployment in 2000. G.dmt is the nickname for the standard officially known as ITU-T Recommendation G.992.1.

G.dmt-based ADSL service may use a distributed splitter approach based on simple inline low-pass filters (LPFs) placed at each phone in the customer premises combined with a high-pass filter (HPF) built into the ADSL modem. Figure 3-1 shows what a typical microfilter looks like.

Figure 3-1:
A typical
LPF
microfilter
used with
a G.lite
installation.

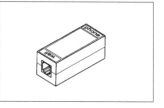

Like all other DSL members, ADSL is distance sensitive. The longer the distance between your premises and the CO, the lower your speed options. Table 3-1 shows the maximum downstream speeds based on selected distances. Keep in mind that actual distance and speed parameters depend on other technical factors.

Table 3-1	ADSL Speeds at Different Distances
Distance of Your Premises from the CO	*Downstream Speed*
Up to 8,000 ft	7 Mbps
Up to 10,000 ft	5 Mbps
Up to 12,000 ft	3 Mbps
Up to 16,000 ft	1 Mbps

G.lite

G.lite is a simplified version of G.dmt ADSL developed for the mass market. G.lite is a nickname for the standard officially known as ITU-T Recommendation G.992.2. It supports 1.5 Mbps downstream and 512 Kbps upstream. G.lite is expected to be widely adopted by consumers. This standards-based DSL is easier and less expensive for DSL providers to install and promises to bring down the cost of DSL modems and service for DSL customers. This means G.lite CPE will soon be available in the same way dial-up modems are today: through retail stores and other sales channels. G.lite, like ADSL, delivers simultaneous high-speed data and voice service over the same telephone line.

Although G.lite is referred to as splitterless ADSL, it may use a distributed splitter approach based on simple inline low-pass filters (LPFs) placed at each phone in the customer premises combined with a high-pass filter (HPF) built into the ADSL modem. These microfilters may be used to address potentially disturbing POTS devices (telephones, faxes, and modems) that can cause interference for a DSL connection. Figure 3-2 shows how G.lite would be configured using LPFs. Current ADSL service may also use these same filters.

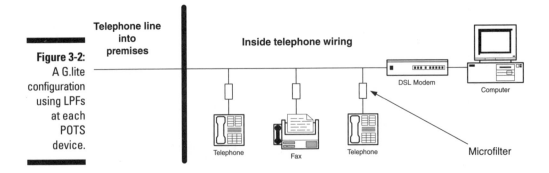

Figure 3-2: A G.lite configuration using LPFs at each POTS device.

G.lite is an international standard for ADSL (G.992.2) It uses DMT line-coding technology, which is based on the U.S. ANSI ADSL standard, T1.413 Issue 2. At the time of this writing, two unknowns about G.lite are the role of microfilters and the impact of fast-retrain (a feature unique to G.lite that makes it a more robust, consumer-installable solution).

SDSL

Symmetric Digital Subscriber Line (SDSL), as its name implies, is a symmetric service that can deliver variable high-speed data communication speeds up to 2.3 Mbps. SDSL can be packaged in a range of bandwidth configurations that include 160 Kbps, 192 Kbps, 384 Kbps, 416 Kbps, 768 Kbps, 784 Kbps, 1.04 Mbps, 1.1 Mbps, 1.5 Mbps, and 2.3 Mbps, depending on the DSL provider.

SDSL, unlike ADSL, doesn't support the use of analog on the same line. SDSL uses 2B1Q line coding, which is widely used by the telephone companies for such services as ISDN lines. SDSL and HDSL are related members of the DSL family. HDSL is a proven DSL technology that has been used in T1 lines for years.

SDSL service can't be used concurrently over an existing POTS line. A separate telephone line is installed for the SDSL service. However, the deployment of *Voice over DSL (VoDSL)* technology will allow voice communications to be handled over an SDSL.

Symmetric transmission gives you faster upstream speed, which is one reason businesses find SDSL attractive. Leading CLECs, such as NorthPoint Communications and Covad Communications, are aggressively deploying SDSL. Table 3-2 shows the speed parameters for SDSL service based on the distance between the customer's premises and the CO.

Table 3-2	SDSL Distance and Speed Parameters
Distance of Your Premises from the CO	*Speed*
Up to 9,500 ft	1.5 Mbps
Up to 12,500 ft	1.04 Mbps
Up to 14,900 ft	784 Kbps
Up to 18,000 ft	416 Kbps
Up to 19,000 ft	384 Kbps
Up to 20,000 ft	200 Kbps
Up to 22,770 ft	160 Kbps

G.shdsl is an emerging standard for SDSL. G.shdsl is a rate-adaptive variant of HDSL-2 and promises to become the replacement for the current SDSL. It provides a longer loop reach while being more friendly to other services that may be deployed within the binder.

IDSL

ISDN Digital Subscriber Line (IDSL) is the always-on cousin of dial-up ISDN (Integrated Services Digital Network). IDSL delivers up to 144 Kbps of symmetric bandwidth, which is 16 Kbps more than the dial-up version of ISDN. This 16-Kbps difference comes from the elimination of the two 8-Kbps channels used in ISDN for signaling the PSTN switch. IDSL, like ISDN, uses the 2BIQ line-coding scheme and doesn't support analog communications.

IDSL is attractive because it has a much larger range than the other DSL flavors, reaching up to 36,000 feet. IDSL has another benefit; it can be deployed through DLCs (Digital Loop Carriers).

IDSL offers a more affordable alternative for many dial-up ISDN users facing high usage costs when they use more than 20 or 30 hours per month. Another advantage to making the ISDN-to-IDSL conversion is that you can use existing ISDN CPE with IDSL.

The downside of IDSL is that it's not scalable, which means you can't upgrade your speed. ILECs and CLECs are both offering IDSL services to address the larger potential market than the one available with their other DSL services alone.

HDSL and HDSL-2

High-bit-rate Digital Subscriber Line (HDSL) is the most mature DSL technology. HDSL has been used by ILECs for years as an alternative to traditional T1 lines, so it's often sold at a high cost. HDSL provides symmetric data communications of up to 1.54 Mbps. HDSL requires the use of two pairs (four wires) instead of the standard single pair used for other DSL flavors and doesn't support POTS.

The U.S DSL standards body, ANSI T1, has approved a new generation of HDSL called *HDSL-2* that offers several enhancements over its predecessor. One of the most important is that HDSL-2 requires only a single twisted-pair local loop instead of the two pairs required for HDSL. HDSL-2 delivers its data communications service over a single copper-wire pair to a distance of up to 18,000 feet without the use of repeaters.

VDSL

Very-high-bit-rate Digital Subscriber Line (VDSL) is the ultra-high-speed DSL flavor. VDSL is a technology capable of supporting symmetric or asymmetric services and can be used by premises in close proximity to a central office of DLS. The closer the customer is to the CO, the higher the speed capabilities. For example, at 1,000 feet from the CO, the downstream rate can reach up to 52 Mbps. At 3,000 feet, a subscriber can attain data rates up to 26 Mbps, and at about 5,000 feet, data rates can reach up to 13 Mbps. VDSL can provide asymmetric data and POTS transmissions on a single twisted pair.

MVL

Hotwire *Multiple Virtual Lines (MVL)* is a proprietary version of DSL developed by Paradyne. Up to eight MVL modems can be connected to a single POTS wire pair, enabling each to have a bandwidth of up to 768 Kbps. MVL is a splitterless technology, so service technicians are not required at the customer's premises.

Hotwire MVL is used by a small number of DSL service providers, such as Choice One Communications, HarvardNet, Network Access Solutions, and Rhythms. Chapter 4 provides more information on DSL providers.

In Search of DSL Standards

Standards are agreed principles of protocol set by committees working with various industry and international standards organizations. The standards process works through consensus. Experts from across an industry meet, debate, and share options and research to arrive at a standard. Following are the big benefits consumers gain from standards:

- **Standards fuel competition** among CPE manufacturers and data communication service providers. This lowers costs while providing added value to the consumer as vendors look to differentiate their products by adding new features.

- **Standards usually create a solution that is greater than the sum of its parts.** The process of different experts from a variety of companies working together and discussing the technology leads to a solution significantly better than one company could develop on its own. Note, however, that standards are compromises of great competing working solutions.

- **Standards lead to interoperability among devices,** so consumers can buy products from multiple vendors.

✔ **Standards enable mass-deployment.** By enabling interoperability, consumers can purchase CPE in confidence knowing it will work.

✔ **Standards create a level playing field** that enables service providers and product vendors to offer DSL CPE and services at the lowest possible prices due to competition.

✔ **Standards focus the industry on resolving other challenges to deployment**. When interoperable, standards-based products are widely available, the industry collectively focuses on building the standards-based infrastructure to support DSL deployment.

Know your DSL standards organizations

Two major standards bodies influence DSL:

✔ The **International Telecommunication Union (ITU)** is an agency of the United Nations that is the primary source of telecommunications standards. You can check out the ITU on the Web at `www.itu.int`.

✔ The **American National Standards Institute (ANSI)** is the primary standards facilitating organization in the United States. ANSI facilitates the development of standards by establishing consensus among qualified groups. You can check out ANSI on the Web at `www.ansi.org`. The group that standardized DSL is ANSI chartered ATIS T1E1.4, a working group within ATIS T1. ATIS stands for Alliance for Telecommunications Industry Solutions. You can check them out at `www.atis.org`.

The leading DSL industry organization is the DSL Forum (formally the ADSL Forum). The DSL Forum represents most telecommunications and computer companies involved with DSL services and is working to develop consensus on DSL standards. You can check out the DSL Forum on the Web at `www.dslforum.org`. The DSL Forum also maintains a consumer education site at DSL Life (`www.dsllife.com`).

The current state of DSL standards

The current state of DSL standards is one of steady progress. Although universal DSL CPE interoperability is still a work-in-progress, significant standards developments have occurred. The G.dmt and G.lite standards for ADSL and G.shdsl are two important standards that will have an effect on consumer DSL and business DSL service, respectively.

At the time of this writing, the next generation of SDSL, referred to as G.shdsl, is working its way towards ITU-T standardization. G.shdsl is a rate-adaptive variant of HDSL-2 and promises to replace the current SDSL.

Chapter 4

Checking Out DSL Providers

. . . .

In This Chapter

▶ Getting DSL service from a data or packet CLEC

▶ Checking out the leading CLEC DSL deployments

▶ Finding ISPs offering CLEC-based DSL circuit service

▶ DSL service offerings from ILECs

. . . .

CLECs and ILECs are scrambling to deploy DSL service. This chapter presents an overview of DSL offerings from CLECs and ILECs, but keep in mind that this information is subject to change in the rapidly changing DSL deployment scene.

Is DSL Available in Your Area?

All the talk about how great DSL is means nothing to you if it's not available in your area. DSL's deployment is on a CO-by-CO basis, which results in a patchwork of DSL service availability. The good news is that DSL providers — both CLECs and most ILECs — have been aggressively building out their DSL networks. If you live in a large metropolitan area with high concentrations of potential DSL customers, chances are good that an ILEC, a CLEC, or an ISP acting as a DSL CLEC is offering DSL service in your area. Many smaller metropolitan areas as well as many towns and rural areas are getting DSL deployments by CLECs and ILECs.

DSL service is part of a total Internet service package, so you need to evaluate the Internet access service being bundled with the DSL service. For example, many lower cost DSL service offerings restrict the types of applications you can use or the number of computers you can connect to the DSL line.

Tracking down DSL service

The Web is the number one resource for checking the availability of DSL service, and the best starting points are ILEC and CLEC Web sites. Most of these sites include Web forms to check DSL availability based on your telephone number. Most DSL providers also list their ISP partners at their Web sites, so you can check out their DSL service offerings as packaged by their ISP partners. Most ILECs also operate their own in-house ISPs.

If you know ISPs in your areas, you can check their Web sites or call them to find out whether they're offering DSL. Depending on the ISP, they might offer DSL service from multiple DSL providers or exclusively from a single DSL provider. Because different DSL providers offer different types of DSL service, you want to compare the different services.

A good Web site for checking out ISPs offering DSL service is DSLReports.com at www.dslreports.com. This information-packed site includes a database of ISPs offering DSL service and thousands of subscriber reviews rating the quality of service from ISPs. Another site for checking DSL availability is www.getspeed.com.

Have your phone number and address ready

Checking whether DSL is available is as easy as filling out a Web prequalification form at a Web site, typically the DSL provider's or ISP's site. You typically enter two pieces of information for prequalification. The first is an existing telephone number already in place at the your premises. Specifically, the area code and exchange prefix (the three digits after the area code) hone in on the CO servicing your area. The DSL provider matches this information against the deployment of their DSLAMs. If the DSL provider has a DSLAM installed at the CO servicing your area, you pass the first hurdle. All this tells you, however, is that a DSLAM is installed at the CO.

The second entry is your address, which is used to determine the distance of your premises from the CO, where the DSLAM is located. As you recall, DSL service is distance sensitive. The farther away your premises are from the CO, the fewer speed options are available. If your location is relatively close to the CO, which is the case in many urban areas, you have access to the full range of DSL speeds being offered. If your premises are located at a distance beyond the range of some of the higher speeds, you have fewer speed options. If your premises are located beyond the distance threshold for the DSL service at any speed, you're out of luck.

Although using the combination of your telephone number and address provides a reasonable guess in determining whether you have access to DSL service, a variety of technical reasons might, in the end, mean that DSL won't be in your future. The quality of the copper wiring, interference in the ILEC's cable plant, and other technical issues can put an end to your DSL dreams.

If the DSL provider says DSL service isn't available and you're within 18,000 feet, request a more accurate manual line test. And request the telephone company clean the line by removing any coils on the line, if necessary.

DSL Service from CLECs

Empowered by the Telecommunications Act of 1996, CLECs are offering DSL circuits at affordable prices to ISPs and in some cases directly to subscribers. CLECs are the most aggressive in deploying DSL technologies on a national scale because they are focused exclusively on data applications and therefore deploy DSL more rapidly. They typically offer DSL services through ISP partners in most major metropolitan markets but might also sell directly to larger corporate customers.

The three leading national CLECs are Covad Communications, NorthPoint Communications, and Rhythms NetConnections. These DSL circuit providers sell DSL connections wholesale to ISPs as well as directly to large corporate clients. The first round of DSL deployments by these CLECs were targeted at large metropolitan markets because they represent the largest concentrations of potential DSL customers. In 2000, data CLECs will be offering DSL service to smaller cities.

DSL CLECs have been primarily focused on small to mid-sized business but are rapidly expanding into the consumer market. The main flavor of DSL they're deploying for businesses is SDSL (Symmetrical DSL). CLECs are aggressively entering the consumer DSL market with ADSL and G.lite offerings that take advantage of line sharing.

Because most DSL circuits installed by CLECs are for ISPs, you typically don't have direct contact with the CLEC itself, except during the installation of the DSL service at your premises. The person who does the inside wiring for DSL service and sets up the DSL CPE (Customer Premises Equipment) for your ISP is almost always a CLEC employee or contractor.

The CLEC and ISP relationship

The CLEC and ISP relationship is one of a partnership between a wholesaler and retailer. The CLEC handles the DSL circuit (the telecommunications realm), and the ISP provides Internet access service bundled with DSL service.

CLECs offer DSL circuits to ISPs at a flat rate. Many ISPs offering DSL service resell DSL circuit services from more than one CLEC as well as from ILECs. The cost to the ISP for the DSL connection varies depending on the CLEC, the speed of the service, and other factors. The ISP sells the DSL circuit as part of an Internet service package. ISPs may break down the cost of the DSL circuit and the Internet service, but the prices they show for the DSL circuit aren't necessarily what they are being charged. The ISP might include a markup. You can't know what an ISP is paying for the DSL circuit because it's a negotiated deal.

Because the ISP determines the price for DSL service, prices vary from one ISP to another and from one market to another. The wide variances in DSL service pricing means that you can end up spending hundreds of dollars more per year for the same DSL service from the same CLEC but using a different ISP.

Checking out CLEC DSL availability

Although the ISP is your interface to getting specific information about your DSL connection to the Internet, the CLEC's Web site is the best starting point for checking DSL availability in your area. The CLEC's Web site might also provide links to their ISP partners, which you can then use to check out the ISPs servicing your area to do comparative shopping.

Some CLEC Web sites, however, don't list ISP partners and instead forward to one of their ISP partners the information you entered in checking availability. The partner they choose may or may not be the ISP partner that you ultimately decide to go with for your DSL service. You should always do some comparative shopping when it comes to getting your DSL service.

Your CPE options

Like most DSL providers, the CLEC DSL provider determines the DSL CPE that can be used with their DSL service by virtue of the DSLAM equipment they use at the CO. Most CLEC DSL offerings use bridges (DSL modems) and routers. They support DSL CPE from companies such as 3Com, FlowPoint, Efficient Networks, and Netopia.

The DSL CPE and the DSL circuit are sold through the ISP. And like DSL service pricing, costs of DSL equipment vary from one ISP to another. Most ISPs, however, try to keep CPE prices low because they represent one of the biggest start-up costs in getting DSL service. Pricing for your DSL CPE typically ranges between free (for DSL modems) to $500 for routers. Many ISPs offer free or reduced price DSL modems as part of an installation promotion for a one-year commitment. Typically, the CLEC subsidizes the cost of the CPE and installation for their ISP partners.

Double-check any modem giveaway or free set-up offer to see what the time commitment for the service will be. Some can go as high as three years. You don't want to have the kind of lock-in for DSL service.

Most business DSL offerings bundled with a router are more expensive. Routers include more features than a bridge for managing the Internet connection, including support for VPN (virtual private networking), firewalls, and NAT (Network Address Translation). Routed Internet service usually includes static IP addresses to make every PC on the LAN a host on the Internet.

DSL modems are less expensive and are commonly used with lower cost consumer and SOHO (small office/home office) DSL offerings. Typically, these DSL modem offerings include only one static IP address to restrict the number of computers you can connect to the service.

A growing number of vendors are offering a new generation of easy-to-use and affordable routers that sit behind the DSL modem and provide the same router functionality of a DSL router for considerably less money. You might find that getting a DSL modem and adding your own router behind it is less expensive than going with routed Internet service. Routers that sit behind a DSL bridge let you use the single IP address to support multiple computers connected to a LAN. Chapter 10 explains these router solutions in more detail.

NorthPoint Communications

NorthPoint Communications (www.northpoint.net) is a data CLEC offering DSL service to small and mid-sized businesses and consumers. NorthPoint offers consumer DSL service through ISPs as well as through Radio Shack.

You can use the NorthPoint Web site to check DSL service availability in your area. Or call NorthPoint at 800-981-8213 or 877-836-7375 for more information about their DSL service.

DSL service options

NorthPoint's DSL is priced at a flat rate to ISPs based on the speed of the connection, which is shown in Table 4-1. Your DSL service cost reflects the NorthPoint circuit costs plus the ISP Internet access service charges and ISP markup. NorthPoint brands their DSL service as NorthPoint DSL.

Table 4-1	DSL Service from NorthPoint
Service	*Speed*
IDSL	Up to 144 Kbps/144 Kbps
SDSL	Up to 160 Kbps/160 Kbps
SDSL	Up to 416 Kbps/416 Kbps
SDSL	Up to 784 Kbps/784 Kbps
SDSL	Up to 1.04 Mbps/1.04 Mbps
SDSL	Up to 1.54 Mbps/1.54 Mbps

DSL service availability

The NorthPoint Web site (www.northpoint.net) is a great place for checking whether you can get DSL service. The site displays a national map of metropolitan areas that NorthPoint serves. You can fill out a prequalification form to check whether NorthPoint DSL service is available in your area. The NorthPoint Web site currently includes links to the Web sites of NorthPoint's ISP partners.

NorthPoint DSL service is available in the following metropolitan areas (which are alphabetized according to state):

> Phoenix; Los Angeles, Sacramento, San Diego, and San Francisco; Denver; Washington, D.C.; Ft. Lauderdale, Miami, Orlando, St. Petersburg Tampa; Atlanta; Chicago; Indianapolis; Kansas City; Baltimore; Boston; Detroit; Minneapolis, St. Paul; St. Louis; New York City; Cleveland and Columbus; Portland (Oregon); Philadelphia and Pittsburgh; Providence; Raleigh/ Durham; Austin, Dallas, Houston, and San Antonio; Seattle; Milwaukee;

By the end of 2000, NorthPoint Communications plans to extend its national DSL network to more than 50 percent of all businesses and nearly 45 percent of all homes in the United States. The following new markets (alphabetized by state) will added to the NorthPoint network:

> Birmingham; Tucson; Fresno, Sacramento, and Santa Barbara; Hartford; Jacksonville; Kansas City and Wichita; Louisville; New Orleans; Grand Rapids; Las Vegas; Albuquerque; Albany, Buffalo, Rochester, and Syracuse; Charlotte and Greensboro; Cincinnati, Columbus, and Dayton; Oklahoma City; Harrisburg; Providence; Memphis and Nashville; San Antonio; Salt Lake City; Norfolk and Richmond

CPE options

NorthPoint uses Copper Mountain (www.coppermountain.com) as its DSLAM equipment vendor, which in turn defines what DSL CPE can be used by NorthPoint. Each ISP has its own CPE preferences, and many support only a subset of the available CPE options supported by Copper Mountain DSLAMs. Currently, NorthPoint Communications supports the following DSL CPE:

- ✔ FlowPoint 2200 SDSL router
- ✔ FlowPoint 144 IDSL router
- ✔ Netopia M-7100 SDSL modem
- ✔ Netopia R-7100 SDSL router
- ✔ 3Com SDSL modem
- ✔ 3Com Office Connect 840 SDSL router
- ✔ XPEED 320 SDSL Ethernet modem

Covad Communications

Covad Communications (www.covad.com) is the leading CLEC. Covad has focused on delivering business-class DSL services to small and mid-sized businesses and Internet power users, but they are also aggressively rolling out consumer DSL service. You can call Covad for more information about their DSL service offerings at 800-462-6823.

DSL service options

Covad offers SDSL, ADSL, and IDSL services under the TeleSpeed service mark. Table 4-2 lists Covad's current TeleSpeed services.

Table 4-2	DSL Service from Covad
Service	*Speed*
TeleSpeed 144 (IDSL)	Up to 144 Kbps/144 Kbps
TeleSpeed 192 (SDSL)	Up to 192 Kbps/192 Kbps
TeleSpeed 384 (SDSL)	Up to 384 Kbps/384 Kbps
TeleSpeed 768 (SDSL)	Up to 768 Kbps/768 Kbps
TeleSpeed 1.1 (SDSL)	Up to 1.1 Mbps/1.1 Mbps
TeleSpeed 1.5 (ADSL)	Up to 1.5 Mbps/384 Kbps

DSL service availability

At Covad's Web site, you check DSL service availability in your area by completing a series of online forms. Then the site makes a recommendation on the best service option. At the end of the process, Covad supplies links to ISP partners serving your area.

Covad is offering DSL service in the following metropolitan areas (organized by state):

> Phoenix; Los Angeles, Sacramento, San Diego, and San Francisco; Denver; Washington, D.C.; Miami; Atlanta; Chicago; Baltimore; Boston; Detroit; Minneapolis; New York City; Raleigh; Portland (Oregon); Philadelphia; Austin, Dallas, and Houston; Seattle

In 2000, Covad plans to add the following metropolitan areas (in state order):

> Birmingham; Colorado Springs; Hartford; Jacksonville; Indianapolis; Grand Rapids; Kansas City; Las Vegas; Albuquerque; Albany and Buffalo; Charlotte and Greensboro; Akron, Cincinnati, Cleveland, Columbus, and Dayton; Allentown and Harrisburg; Charleston

CPE options

Covad uses Nokia DSLAMs, which currently support the following DSL CPE :

- ✔ Efficient Networks SpeedStream 5250 SDSL bridge
- ✔ FlowPoint 144 IDSL router
- ✔ FlowPoint 2200 SDSL router
- ✔ Netopia R-7200 SDSL router

Rhythms NetConnections

Rhythms NetConnections (www.rhythms.net) is focused primarily on working with large companies to support teleworkers and branch offices. They also provide DSL service to ISP partners, which are listed on the Rhythms Web site.

DSL service availability

You can check out DSL availability in your area on the Rhythms Web site. You can also call Rhythms for more information on their DSL service at 800-749-8467.

Rhythms offers DSL service in the following metropolitan areas (alphabetized by state):

Phoenix; Los Angeles, Oakland, Sacramento, San Diego, San Francisco, and San Jose; Denver; Hartford; Washington, D.C.; Miami, Tampa/St. Petersburg; Atlanta; Chicago; Indianapolis; Kansas City; Baltimore; Boston; Minneapolis; Detroit; St. Louis; New York City; Cincinnati, Cleveland, and Columbus; Portland (Oregon); Philadelphia and Pittsburgh; Durham, Raleigh; Austin, Dallas, Fort Worth, Houston, and San Antonio; Salt Lake City; Seattle; Milwaukee

In 2000, Rhythms plans to roll out DSL service in the following markets (by state):

Birmingham; Southern Connecticut; Wilmington; Jacksonville, Tampa/St. Petersburg, and West Palm Beach; Des Moines; Louisville and Memphis; New Orleans; St. Louis; Kansas City; Omaha; Albany, Buffalo, and Rochester; Charlotte and Greensboro; Cincinnati and Dayton; Harrisburg; Providence; Memphis and Nashville; San Antonio; Norfolk and Richmond

CPE options

Rhythms uses DSLAMs from Cisco, Copper Mountain, and Paradyne. The company offers RADSL, SDSL, and IDSL services and supports the following DSL CPE:

- Cisco (www.cisco.com) Cisco 675 and ISDN routers
- FlowPoint (www.flowpoint.com) 144 ISDL router and 2000 SDSL router
- Netopia (www.netopia.com) R7100 SDSL router and R3100 ISDL router
- Paradyne (www.paradyne.com) 5446 and 5446 RADSL bridge and router plus MVL modem

More CLECs popping up everywhere

CLECs offering DSL service are popping up like mushrooms after a spring rain. Many of these players are regional DSL providers; others have national ambitions. Some are ISPs acting as data CLECs; others are data CLECs that wholesale DSL service to ISPs. Finding a comprehensive listing of ISPs acting as their own DSL CLECs is not easy because no single information resource is available. Table 4-3, however, lists some of them.

Table 4-3	Regional CLECs and Emerging National CLECs	
CLEC	*Web Site*	*Service Area*
HarvardNet	www.harvardnet.com	Maine; Massachusetts; New Hampshire; Rhode Island
InterAcess	www.interaccess.com	Chicago area
Jato Communications	www.jato.net	Colorado; Provo and, Salt Lake City; Albuquerque and Santa Fe
Florida Digital Network	www.floridadigital.net/	Orlando, Fort Lauderdale, Jacksonville, West Palm Beach, Miami, Tampa
Network Access Solutions	www.nas-corp.com	Bell Atlantic territory
Bluestar Communications	www.bluestar.com	Birmingham and Huntsville; Little Rock; Boca Raton, Jacksonville, Orlando, and Tampa; Columbus and Decatur; Louisville and Lexington; Baton Rouge, New Orleans, and Shreveport; Jackson; Charlotte, Greensboro, and Raleigh; Charleston and Columbia; Chattanooga, Knoxville, Memphis, and Nashville
New Edge Networks	www.newedgenetworks.com	Washington; Oregon; Idaho; currently certified in 27 states with plans to be certified in all 50 states
Vitts	www.vitts.com	New England area
ChoiceOne Communications	www.choiceonecom.com	Pennsylvania; New York; New Hampshire; Rhode Island; Massachusetts

DSL Service from Your Telephone Company

You know ILECs as the telephone company that handles your voice and analog modem and ISDN connections. This section presents an overview of DSL service being offered by the leading ILECs — Pacific Bell, US WEST, Bell Atlantic, BellSouth, Southwestern Bell, Ameritech, and GTE — as well as some smaller ones, such as Cincinnati Bell, Nevada Bell, and SNET. Because Pacific Bell, Southwestern Bell, Nevada Bell, and SNET are all owned by SBC, their DSL service offerings are similar.

Most ILECs are aggressively deploying DSL services, and the primary DSL flavor being deployed is ADSL. ADSL can be delivered over the same line as your POTS service.

ILECs are offering ADSL offerings through their own in-house ISPs as well as providing DSL circuit services to ISP partners. ILEC ISP subsidiaries have names such as Bell Atlantic.net, PacBell.net, and USWest.net. If you get DSL service through an ILEC ISP partner, your DSL service might be billed separately by the ILEC. If you get DSL service from the ILEC ISP service, all charges are usually included on one phone bill.

ILEC consumer offerings from their in-house ISPs are typically single-PC solutions that use either a single static or dynamic IP address with no support for DNS. You can use an Internet-connection-sharing solution to get around this restriction. See Chapter 10 for more information on sharing a DSL connection.

The ILEC CPE options are typically single-user bridges, PCI DSL modem cards, or USB modems. Some ILECs, such as SBC, also offer business-class DSL services through their in-house ISPs using routers, which includes multiple IP addresses, multiple email boxes, and DNS service.

Many ILECs have improved their Web sites to provide more information for DSL shoppers. Most include prequalification forms to check out availability of DSL service in your area. These sites often include detailed information about the ILEC's DSL service offerings.

Pacific Bell

Pacific Bell, which is owned by SBC Communications, is one of the leaders in ILEC deployments of DSL service. PacBell is aggressively rolling out DSL service, driven by the fact that an estimated 35 percent of the nation's Internet traffic begins and ends in California.

Pacific Bell uses DSLAMs made by Alcatel, the giant French telecommunications company. PacBell offers DSL CPE from Alcatel, Westell, and others.

DSL service options

Pacific Bell offers a range of DSL Internet service packages from PacBell.net. For details, check out Table 4-4.

Table 4-4		DSL Service from PacBell	
Service	*Speed*	*Cost*	*One-Time Fees*
Basic DSL	384 Kbps–1.54 Mbps/128 Kbps	$39 a month for Pacific Bell ADSL and Internet access (based on a 1-year commitment). This is a dynamic IP address Internet access account that does not allow any commercial or business servers to operate on the DSL connection. The DSL equipment is free (with a 1-year commitment).	$100 installation charge. $200 fee applies for early termination of the 1-year contract.
Enhanced DSL	384 Kbps–1.54 Mbps/128 Kbps	$79 a month (based on a 1-year commitment) with Internet access. Includes five usable IP addresses and DNS service. The DSL equipment is free (with a 1-year commitment).	$100 service installation charge. $200 fee applies for early termination of the 1-year contract.
	1.54 Mbps–6 Mbps/384 Kbps	$199 a month (based on a 1-year commitment) with Internet access. Includes five usable IP addresses and DNS service. The DSL equipment is free (with a 1-year commitment).	$100 service installation charge. $200 fee applies for early termination of the 1-year contract.

Service	Speed	Cost	One-Time Fees
Business DSL	384 Kbps–1.54 Mbps/128 Kbps	$238 a month (based on a 1-year commitment) with Internet access. Includes 29 usable IP addresses and DNS service. The DSL equipment is free (with a 1-year commitment).	$200 service installation charge. $200 fee applies for early termination of the 1-year contract.
	1.54 Mbps–6 Mbps/384 Kbps	$328 a month (based on a 1-year commitment) with Internet access. Includes 29 usable IP addresses and DNS service. The DSL equipment is free (with a 1-year commitment).	$200 service installation charge. $200 fee applies for early termination of the 1-year contract.

To get the latest information on the PacBell DSL offerings, including in formation on their ISP partners, check out the Pacific Bell Web site at www.pacbell.com.

You can also call the FasTrak DSL Center at 1-888-884-2DSL (2375) Monday through Friday, 8 A.M. to 5 P.M.

DSL service availability

Pacific Bell offers ADSL service in many California communities. Check their Web site to see whether service is available in your area.

US WEST

US WEST was out of the DSL starting gate early with a range of service options. MegaBit Services is the umbrella name for US WEST's family of DSL services.

DSL service options

Table 4-5 describes the US WEST DSL service offerings based on a one-year contract. Pricing may change from state to state.

Table 4-5	DSL Service from US WEST		
Service	*Speed*	*Cost*	*One-Time Fees*
MegaBit 256 Select	256 Kbps/256 Kbps dial-up DSL service	$20 a month with no Internet service	Connection fee of $69 plus a Cisco modem ($299) or PCI card ($199)
MegaBit 256 Deluxe	256 Kbps/256 Kbps	$30 a month with no Internet service	Connection fee of $69 plus a Cisco 675 external modem ($295) and the Cisco 605 internal modem card ($145) or the Intel 2100 internal modem card ($145)
MegaOffice	512 Kbps/512 Kbps	$62 a month with no Internet service	Connection fee of $69 plus a Cisco 675 external modem ($295) and the Cisco 605 internal modem card ($145) or the Intel 2100 internal modem card ($145)
MegaBusiness	768 Kbps/768 Kbps	$80 a month with no Internet service	Connection fee of $69 plus a Cisco 675 external modem ($295) and the Cisco 605 internal modem card ($145) or the Intel 2100 internal modem card ($145)
MegaBit	1 Mbps/1 Mbps	$120 a month with no Internet service	One-time connection fee of $69 plus a Cisco 675 external modem ($295) and the Cisco 605 internal modem card ($145) or the Intel 2100 internal modem card ($145)

Service	Speed	Cost	One-Time Fees
MegaBit	4 Mbps/1 Mbps	$480 a month with no Internet service	Connection fee of $69 plus a Cisco 675 external modem ($295) and the Cisco 605 internal modem card ($145) or the Intel 2100 internal modem card ($145)
MegaBit	7 Mbps/1 Mbps	$840 a month with no Internet service	One-time connection fee of $69 plus a Cisco 675 external modem ($295) and the Cisco 605 internal modem card ($145) or the Intel 2100 internal modem card ($145)
MegaBit 256 Deluxe	256 Kbps/256 Kbps (dial-up DSL service) including Internet service	$37 a month	Connection fee of $74 plus a Cisco 675 external modem ($295) and the Cisco 605 internal modem card ($145) or the Intel 2100 internal modem card ($145)
MegaBit 256 Deluxe	256 Kbps/256 Kbps including Internet service	$47 a month	Connection fee of $74 plus a Cisco 675 external modem ($295) and the Cisco 605 internal modem card ($145) or the Intel 2100 internal modem card ($145)

Customers are assessed a one-time installation charge and a monthly charge. You must also purchase a DSL modem.

The normal installation charge is $69 for a customer self-install, but you can get an on-site modem and software installation from US West for $150. You can check out US WEST DSL offerings on the Web at www.uswest.com/products/data/dsl or call 877-665-6342.

DSL service availability

US WEST is currently offering DSL service and partnering with ISPs in the states and communities listed in the following. US WEST will be offering DSL service in New Mexico in 2000, the only state in the US WEST territory without DSL service from US WEST.

In Arizona: Ahwatukee-Foothills, Apache Junction, Chandler, Deer Valley, Fort McDowell, Fountain Hills, Gilbert, Glendale, Litchfield Park, Mesa, North Phoenix, Paradise Valley, Peoria, Phoenix, Scottsdale, Sun City, Tempe, Tolleson, Tucson

In Colorado: Arvada, Aurora, Broomfield, Boulder, Castle Rock, Colorado Springs, Denver, Englewood, Fort Collins, Greeley, Lakewood, Littleton, Longmont, Table Mesa, Vail

In Idaho: Boise, Meridian, Nampa

In Iowa: Ames, Ankeny, Bettendorf, Blue Grass, Cedar Rapids, Council Bluffs, Des Moines, Davenport, Muscatine

In Minnesota: Anoka, Blaine, Bloomington, Brooklyn Center, Burnsville, Coon Rapids, Cottage Grove, Crystal Eagan, East Grand Forks, Eden Praire, Fridley, Golden Valley, Hopkins, Maplewood, Minneapolis, New Brighton, North St. Paul, Plymouth, Shoreview, St. Cloud, St. Paul, Rochester, Wayzata, West St. Paul, White Bear Lake

In Montana: Helena

In Nebraska: Bellevue, Omaha

In North Dakota: Bismarck, Fargo, Grand Forks, Moorhead, West Fargo

In Oregon: Albany, Albany, Corvallis, Eugene, Lake Oswego, Milwaukee-Oak Grove, Oregon City, Portland, Salem, Springfield

In South Dakota: Sioux Falls

In Utah: Bountiful, Clearfield, Farmington, Holladay, Kaysville, Kearns, Murray, Orem, Provo, Salt Lake City

In Washington: Auburn, Bellevue, Bellingham, Des Moines, Federal Way, Issaquah, Kent, Lacey, Mercer Island, Olympia, Puyallup, Renton, Seattle, Spokane, Tacoma, Vancouver

In Wyoming: Cheyenne

US WEST includes a listing of ISP partners offering DSL service using US WEST DSL service. The US WEST portion of your bill is added to your telephone bill. You can check out this list at www.uswest.com/products/data/dsl/isplist.html.

Bell Atlantic

Bell Atlantic is aggressively rolling out ADSL services.

Bell Atlantic uses DSLAMs from Alcatel but uses DSL modems from Westell (www.westell.com), Efficient Networks (www.efficient.com), and 3Com (www.3com.com).

Bell Atlantic plans to market its InfoSpeed service through the retail channel by offering Home Connection Kits in Staples, CompUSA, and Best Buy. These self-installation kits include a DSL modem plus microfilters for connecting to POTS devices in the home. Bell Atlantic and America Online (AOL) have a strategic alliance whereby Bell Atlantic provides DSL service for AOL customers.

DSL service options

Bell Atlantic is offering ADSL service under the InfoSpeed brand name. Bell Atlantic also sells DSL transport services to ISPs. Check out the Bell Atlantic InfoSpeed Web site at www.bell-atl.com/infospeed. You can also check out the Bell Atlantic small business site at www.bell-atl.com/smallbiz or call Bell Atlantic at 877-525-2375 for more information.

Bell Atlantic InfoSpeed service offerings are listed in Table 4-6. Current Bell Atlantic DSL product offerings through Bell Atlantic.net are single-computer DSL consumer solutions, with dynamic IP addressing and no DNS service. The Personal InfoSpeed and Professional InfoSpeed DSL offerings support only 90 Kbps upstream, which is one of the lowest upstream DSL offerings around. Bell Atlantic also offers an upgrade deal for its ISDN customers.

Table 4-6	DSL Service from Bell Atlantic		
Service	*Speed*	*Cost*	*One-Time Fees*
Personal InfoSpeed	640 Kbps/90 Kbps	$50 a month with Internet service; $40 without	Connection fee of $99 including a DSL modem.
Professional InfoSpeed	1.54 Mbps/90 Kbps	$100 a month with Internet service; $60 without	Connection fee of $99 including a DSL modem.
Power InfoSpeed	7.1 Mbps/680 Kbps	$190 a month with Internet service	Connection fee of $99 including a DSL modem.

Bell Atlantic plans a number of improvements to its DSL service in 2000, including

- ✔ Higher upstream rates on their Personal and Professional InfoSpeed offerings.
- ✔ Symmetrical ADSL offerings, which means the upstream and downstream rates will be equal.
- ✔ Routed DSL Internet service offerings with multiple static IP addresses.

DSL service availability

Bell Atlantic provides telecommunication services to the following states:

Maine, New Hampshire, Vermont, Massachusetts, Rhode Island, New York, New Jersey, Pennsylvania, Delaware, Maryland, Virginia, West Virginia, Washington D.C., and a small area in Connecticut.

Following is the current Bell Atlantic DSL deployments by states and communities.

In Maryland: Baltimore, Bethesda, Beltsville, Colesville, Hyattsville, Landover, Rockville, Silver Spring, Suitland, Wheaton

In Massachusetts: Available in 141 communities throughout

In New Jersey: Cliffside Park, Elizabeth, Englewood, Hackensack, Hoboken, Jersey City, Leonia, Newark, North Bergen, Oradell, Rutherford, Union City

In Pennsylvania: Pittsburgh, Philadelphia, Harrisburg

In Virginia: Alexandria, Annandale, Arlington, Baileys Crossroads, Falls Church, Merrifield, Vienna, Richmond

In Washington, D.C.: Dupont Circle, Georgetown, Northwest D.C.

In New York: Albany, Buffalo, Syracuse.

In West Virginia: Charleston and Huntington.

In 2000, DSL service from Bell Atlantic will be available also in New Hampshire, Maine, Rhode Island, and Vermont.

BellSouth

BellSouth is expanding its DSL deployment in its service territory of nine states: Alabama, Florida, Georgia, Kentucky, Louisiana, Mississippi, North Carolina, South Carolina, and Tennessee.

DSL service options

BellSouth DSL service is called FastAccess Internet Service through its ISP subsidiary, which appears to be the only way BellSouth sells its ADSL service. BellSouth is using the Cayman (www.cayman.com) 3220H ADSL modem/router.

BellSouth is offering a 1.54-Mbps/256-Kbps link for $50 per month if you are a subscriber to the BellSouth Complete Choice or Business Choice telephone package. The cost for BellSouth FastAccess service as a standalone product is $60. Both offerings include Internet access from BellSouth.net as well as the ADSL circuit.

Customers are charged a one-time installation fee of $200 for configuring their computer and phone line for FastAccess, and a Service Activation fee of $100, and the cost of the DSL modem, the price of which BellSouth doesn't disclose on their Web site. Check with BellSouth for any special promotions.

You can check out BellSouth's ADSL offerings at two Web sites. Unfortunately, neither site provides information about the specifics of their DSL Internet service and CPE packages, such as IP addresses, DNS, and email. The consumer DSL site is at consumer.bellsouth.net/adsl. The small business DSL site is at dsl.smlbiz.bellsouth.com.

DSL service availability

BellSouth is offering ADSL service in the following areas (alphabetized by state):

> Birmingham, Huntsville, and Montgomery; Boca Raton, Fort Lauderdale, Gainesville, Jacksonville, Miami, Orlando, and West Palm Beach; Athens, Atlanta, and Augusta; Louisville; Baton Rouge, Lafayette, and New Orleans; Jackson; Charlotte, Greensboro, and Raleigh; Charleston, Columbia, and Greenville; Chattanooga, Knoxville, Memphis, and Nashville

Southwestern Bell

Southwestern Bell, which is owned by SBC Communications, is aggressively deploying DSL.

DSL service options

Southwestern Bell ADSL offerings are similar to Pacific Bell (both companies are owned by the same parent company, SBC Communications). These offerings include the ones in Table 4-7.

Table 4-7		DSL Service from Southwestern Bell	
Service	*Speed*	*Cost*	*One-Time Fees*
Basic DSL	384 Kbps–1.54 Mbps/128 Kbps	$39 a month for ADSL and Internet access (based on a 1-year commitment). This is a dynamic IP address Internet access. The DSL equipment is free (with a 1-year commitment).	$100 installation charge. $200 fee applies for early termination of the 1-year contract.
Enhanced DSL	384 Kbps–1.54 Mbps/128 Kbps	$79 a month (based on a 1-year commitment) with Internet access. Includes five usable IP addresses and DNS service. The DSL equipment is free (with a 1-year commitment).	$100 service installation charge. $200 fee applies for early termination of the 1-year contract.
	1.54 Mbps–6 Mbps/384 Kbps	$199 a month (based on a 1-year commitment) with Internet access. Includes five usable IP addresses and DNS service. The DSL equipment is free (with a 1-year commitment).	$100 service installation charge. $200 fee applies for early termination of the 1-year contract.
Business DSL	384 Kbps–1.54 Mbps/128 Kbps	$238 a month (based on a 1-year commitment) with Internet access. Includes 29 usable IP addresses and DNS service. The DSL equipment is free (with a 1-year commitment).	$200 service installation charge. $200 fee applies for early termination of the 1-year contract.

Service	Speed	Cost	One-Time Fees
	1.54 Mbps–6 Mbps/384 Kbps	$328 a month (based on a 1-year commitment) with Internet access. Include 29 usable IP addresses and DNS service. The DSL equipment is free (with a 1-year commitment).	$200 service installation charge.$200 fee applies for early termination of the 1-year contract.

For more information about Southwestern Bell's DSL offerings, check out their Web site at www.swbell.com/dsl or call them at 888-SWB-DSL1 (888-792-3751). Southwestern Bell sells its DSL service through its ISP subsidiary or through ISP partners, which you can check out at the Southwestern Bell Web site.

DSL service availability

Following are the FasTrak DSL deployments for Southwestern Bell as of the end of 1999:

Little Rock; Kansas City, Topeka, and Wichita; St. Louis; Oklahoma City and Tulsa; Abilene, Austin, Beaumont, Dallas, El Paso, Fort Worth, Houston Lubbock, and San Antonio

Ameritech

SBC Communications, the parent company of Pacific Bell, Southwestern Bell, and other ILECs, has added Ameritech to its stable. Ameritech's current DSL offerings are different than the other SBC ILEC DSL service packages. Ameritech is aggressively rolling out DSL since being acquired by SBC.

DSL service options

Ameritech.net's high-speed Internet service offers the SpeedPath packages listed in Table 4-8.

Table 4-8	DSL Service from Ameritech		
Service	*Speed*	*Cost*	*One-Time Fees*
SpeedPath 768	768 Kbps/128 Kbps	$40 a month (based on a 1-year commitment) for ADSL and Internet access. The DSL modem and installation are free. This is a dynamic IP address Internet access account.	$200 fee applies for early termination of the 1-year contract.
SpeedPath 768 Office	768 Kbps/128 Kbps	$70 a month (based on a 1-year commitment) for ADSL and Internet access. The installation is free, and the DSL router charge is $600. This service uses a dynamic IP address but the DSL CPE router costs $450. The service includes five email accounts and add-on options for DNS support.	$200 fee applies for early termination of the 1-year contract.
SpeedPath 768 OfficePlus	768 Kbps/128 Kbps	$100 a month. The installation is free, and the DSL CPE (router) costs $450. This service includes static IP addresses, five email accounts, and add-on options for DNS support.	$200 fee applies for early termination of the 1-year contract.
SpeedPath 1500 OfficePlus	1.5 Mbps/256 Kbps	$180 a month. The installation is free, and the DSL CPE (router) costs $450. This service includes static IP addresses, five email accounts, and add-on options for DNS support.	$200 fee applies for early termination of the 1-year contract.

Ameritech does not offer its SpeedPath DSL service through any ISP partners. You can get the service only through Ameritech.net (www.ameritech.net). For more information, check out www.ameritech.com/products/data/adsl or call 800-910-4369.

DSL service availability

Currently, Ameritech is offering its High-Speed Internet service only in Ann Arbor and Royal Oak, Michigan, and the Detroit and Chicago areas.

GTE

GTE is an independent ILEC offering telecommunications service in a number of states, including the following:

> California, Florida, Hawaii, Illinois, Indiana, Kentucky, Missouri, North Carolina, Ohio, Oregon, Pennsylvania, Texas, Virginia, Washington, Wisconsin

You can check out GTE DSL services at www.get.com/dsl.

DSL service options

GTE offers ADSL service through ISP partners and also from GTE.net (www.gte.net). Table 4-9 lists GTE's DSL service offerings without Internet access service. GTE DSL service uses an Orckit/Fujitsu bridge.

Table 4-9		DSL Service from GTE	
Service	*Speed*	*Cost*	*One-Time Fees*
DSL Bronze Plus	Up to 768 Kbps/ up to 128 Kbps	$50 a month with Internet service from GTE.net and a 1-year contract.	$60 connection fee, $80 inside wiring and modem installation charge, and $199 modem purchase or a modem rental for $12 a month.
DSL Silver	384 Kbps / 384 Kbps	$93 a month with Internet service from GTE.net and a 1-year contract.	$60 connection fee, $80 inside wiring and modem installation charge, and $199 modem purchase or a modem rental for $12 a month.

(continued)

Service	Speed	Cost	One-Time Fees
Table 4-9 *(continued)*			
DSL Gold	768 Kbps / 768 Kbps	$128 a month with Internet service from GTE.net and a 1-year contract.	$60 connection fee, $80 inside wiring and modem installation charge, and $199 modem purchase or a modem rental for $12 a month.
DSL Platinum	1.5 Mbps/ 768 Kbps	$205 a month with Internet service from GTE.net and a 1-year contract.	$60 connection fee, $80 inside wiring and modem installation charge, and $199 modem purchase or a modem rental for $12 a month.
DSL Platinum Plus	1.5 Mbps/ 768 Kbps	Multiuser support. $405 a month with Internet service from GTE.net and a 1-year contract.	$60 connection fee, $80 inside wiring and modem installation charge, and $199 modem purchase or a modem rental for $12 a month.

DSL service availability

GTE is rolling out DSL in select cities across the United States. The following lists GTE DSL deployments:

> Los Angeles, Palm Springs, Santa Barbara, and Santa Monica; Tampa; Bloomington; Elkhart, Fort Wayne, Lafayette, and Terre Haute; Lexington; Columbia; Durham; Beaverton; Erie and York; Carrollton, College Station, Coppell, Dallas, Denton, Flower Mound, Garland, Grapevine, Irving, Lewisville, Plano, San Angelo, and Texarkana; Manassas and Dale City; Seattle, Everett, Redmond, and Wenatachee; Wausau

DSL Internet access must be obtained separately through one of the participating ISPs or from GTE.net. You can check out the GTE list of ISP partners at `www.gte.com/dsl/partisp.html`.

Other Telephone Companies Offering DSL

DSL service is also being offered by some of the smaller ILECs, including Cincinnati Bell, Nevada Bell, SNET, and Sprint.

Cincinnati Bell

Cincinnati Bell Telephone provides telecommunication services in a three-state area consisting of portions of Ohio, Kentucky, and Indiana. Cincinnati Bell offers an ADSL service called ZoomTown, which comes in three different service offerings, as shown in Table 4-10.

Table 4-10		DSL Service from Cincinnati Bell	
Service	*Speed*	*Cost*	*One-Time Fees*
ZoomTown	768 Kbps/384 Kbps	ADSL Internet service package for $40 a month. Internet access is provided by Cincinnati Bell.	$99 installation fee.
ZoomTown TurboSpeed	768 Kbps/384 Kbps	ADSL service for $30 a month without Internet access service. Service is sold through Cincinnati Bell ISP partners, which add Internet service charges on top of the $30 a month.	$99 installation fee.
ZoomTown HyperSpeed	1.5 Mbps/768 Kbps	ADSL service for $160 a month without Internet service. Service is sold through Cincinnati Bell ISP partners, which add Internet service charges on top of the $160 a month.	$99 installation fee.

The ZoomTown service includes four static IP addresses to enable up to four computers to share the connection. You can check out the Cincinnati Bell ZoomTown offerings at www.cinbelltel.com or by calling 513-566-9666.

SNET

SNET, acquired by SBC, provides telecommunication services in Connecticut. Because SNET is part of SBC, its DSL offerings reflect the standard SBC offerings (see Pacific Bell or Southwestern Bell). SNET offers ADSL service in the following Connecticut towns:

> Bridgeport, Bridgeport North, Bristol, Danbury, Meriden, New Haven, Norwich, Stamford, Stamford North, Stratford, Waterbury, Westville.

SNET plans to offer DSL service in the following Connecticut towns by early 2000:

> Easthaven, Hamden, Hartford, Manchester, Milford, Norwalk, Norwalk North, Torrington, Trumbull, Wallingford, Wethersfield.

You can get more information on SNET's DSL offerings on the Web at `www.snet.com/adsl` or by calling 877-999-9DSL (9375).

Nevada Bell

Nevada Bell, owned by SBC, serves roughly 30 percent of the state of Nevada, including the Reno/Sparks metropolitan area and widespread rural areas. Because SNET is part of SBC, its DSL offerings reflect the standard SBC offerings (see Pacific Bell or Southwestern Bell).

You can check out Nevada Bell DSL offerings at `public.nvbell.net/dedicated/dsl` or by calling the DSL Ordering Center at 775-323-2375.

Sprint

Sprint is offering an ADSL and voice communication package called ION. This package includes up to four telephone numbers and high-speed Internet access that all share the single ADSL telephone line. It uses VoDSL technology to deliver the voice communication service. The cost is $160 per month plus $250 for the ION hub service and $150 for installation. It is currently available in Denver, Kansas City, and Seattle.

Contact Sprint ION for more information at 877-746-8466 or on the Web at `www.sprint.com`.

Chapter 5

Bringing DSL to Life with Internet Access

· ·

In This Chapter

▶ Understanding the DSL and ISP connection

▶ Going for bridged or routed Internet service

▶ Figuring out your IP addresses and DNS service needs

▶ Hosting your Internet email, network news, and Web services

· ·

*T*he DSL circuit is the pipeline for high-speed connectivity. What brings the DSL link to life are the value-added services of the Internet service provider. The ISP bundles the DSL circuit service with the IP addresses, Domain Name System services, virtual private networking, and other TCP/IP-based services to define your Internet service. This chapter takes you through the key elements that make up your Internet access package.

Buying Your DSL Service from an ISP

Most DSL service is sold as part of an Internet access package, so your Internet service provider will be our interface to getting a DSL connection to the Internet. As part of your DSL service buying decision, you want to know what IP (Internet Protocol) networking services are included with the service.

What you can and can't do with your DSL Internet connection is defined in large part by the particular configuration of services offered by the ISP.

The United States has thousands of ISPs, and a wide variety of them offer DSL service. As a result, the packaging and pricing of DSL service is all over the map — literally and figuratively.

Three types of ISPs offer DSL-based Internet access service:

- ✔ **Independent ISPs.** These ISPs buy their DSL circuits from CLECs or ILECs and provide the bulk of DSL service offerings. In large metropolitan markets, independent ISPs typically buy DSL service from multiple DSL circuit providers. For example, an ISP serving northern California might offer DSL service from Pacific Bell, NorthPoint, and Covad Communications.

- ✔ **ILEC ISPs.** These ILEC-owned ISPs provide an Internet access package added to the ILEC's DSL offerings. ILEC ISPs have names such as PacBell.net, BellAtlantic.net, and USWEST.net. They compete with independent ISPs but offer only the ILEC's DSL service.

- ✔ **ISPs acting as CLECs**. These ISPs become CLECs by filing tariffs with the state regulatory agency, installing DSLAMs in COs, and using the ILECs' local loops in a similar way that larger CLECs do. A pioneer of this approach is Harvard Net, which is located in the Boston area.

Given the complexity of TCP/IP networking combined with DSL service issues, most ISPs could do a better job of creating user-friendly packaging of DSL-based Internet access services. ISP Web sites are the main source of initial customer information, yet many of these sites lack helpful information for customers trying to define their needs. In many cases, you have to call the ISP for basic service information. And even when you call, you might get a salesperson ill prepared to answer your questions. One site to check out is DSL Reports (www.dslreports.com), which provides DSL customer reviews of ISPs offering DSL service.

Bridged or Routed DSL Service?

DSL CPE (consumer premises equipment) is intertwined with IP networking, which forms the basis of the Internet. DSL CPE comes in two flavors: bridges and routers. Bridges and routers are internetworking devices that allow separate networks to communicate with each other. A *bridge,* commonly called a *DSL modem,* is a device that combines one or more networks into a single seamless network. A *router* is a device that forwards data traffic between separate networks. See Chapter 8 for more information on DSL CPE solutions.

The types and brands of DSL CPE available from a particular ISP is determined by the DSLAM used by the DSL provider whose service the ISP is offering customers. The ISP selling the service sells the CPE as part of the DSL Internet service package.

The most common way DSL CPE connects to a computer is Ethernet, which forms the basis of most computer networking. Ethernet-based DSL CPE connects through an NIC (network interface card) or Ethernet adapter installed in the computer. DSL modems and routers using Ethernet can connect to a single computer or to multiple computers connected to a LAN (local area network). When you use an Ethernet-based DSL CPE to connect to the Internet, all TCP/IP data from your Web browser, email program, or any Internet application passes out to the Internet through the network adapter in your computer. Chapter 11 explains Ethernet and local area networks.

Some Ethernet-based DSL modems are designed to work with a single computer by recognizing TCP/IP data traffic from only one Ethernet adapter.

Single-computer DSL CPE consists of external USB modems and internal PCI adapter cards. The new USB modems are easier to connect to your computer but restrict you from sharing the DSL connection.

Bridging your Internet connection

An ISP can offer a *bridged service* that supports multiple computers on a LAN by providing static IP addresses. Or the ISP can restrict the service to a single computer by providing only a single static IP address. However, you can easily bypass this single-IP-address solution by using an Internet-sharing solution such as a proxy server or an Ethernet router. More on Internet-connection-sharing solutions in Chapter 10.

ISPs offer bridged access to the Internet to manage the routing overhead and complexity on their network, which makes it easier for your business. Because bridges are simpler devices, they can be added to the ISP's network without any configuration of the device on your end. Bridges, unlike routers, don't need to be configured with specific IP address information. DSL bridges typically cost a few hundred dollars less than DSL routers. However, bridges don't include the enhanced capabilities of a router.

Bridged service is simple because the computers on your network are configured to be on the same IP subnet as the router at the ISP's side of the DSL circuit. No subnet routing is involved. The bridge doesn't include the NAT and DHCP features built into the DSL router because the bridge doesn't filter data.

In a bridged service configuration, the ISP supplies IP addresses for each computer on your LAN that you want connected to the DSL service. The downside of using a bridged service is that you have no firewall protection unless you purchase a proxy server, an Ethernet router, or an Internet security appliance. A *firewall* controls access to your LAN by using identification information associated with TCP/IP packets to make a decision about whether to allow or deny access. As a bridge connection, you are part of the ISP's network, with nothing protecting your network from their network.

Converting a single PC connection to a LAN connection

Many low-cost DSL offerings are targeted at single-user DSL consumers. For CPE, CLECs and ILECs are increasingly using USB (Universal Serial Bus) and PCI (Peripheral Component Interconnect) adapter card modems but they also use Ethernet DSL modems. They also typically use dynamic IP addresses and don't support the use of DNS.

The ISP's router acts as a DHCP server and dynamically assigns IP addresses temporarily to your DSL CPE with different leasing times. Your IP address changes randomly, so you typically don't have a domain name. This means also that you don't have DNS, so you also can't run any type of Internet server or use IP video and voice applications. You can share these DSL connections, however, by using either an Ethernet router or proxy server software.

A growing nymber of hardware and software solutions convert these single-computer DSL connections to multiuser connections. These products include Ethernet routers (which sits between your LAN and the DSL bridge), proxy servers, and Internet-connection-sharing software. They allow you to share a dynamic IP address or a single static IP address account across multiple computers on a LAN.

Many bridged DSL service packages are targeted at the consumer and SOHO (small office/home office) market. ILECs tend to be the most restrictive in their bridged service offerings. Many use dynamic IP addresses as part of the service. Not having a static IP address can restrict you from effectively using certain types of Internet applications, such as VPN (virtual private networking) and video conferencing. (To find out more, see Chapter 7.)

The best DSL bridged service includes at least one static IP address. You can then leverage that single IP address account to support multiple computers and take advantage of Internet applications such as VPN, video conferencing, or running your own private Web server.

Getting fancy with routed service

With routed service, your LAN is defined as a separate network from your ISP's network. Routed DSL service uses subnet mask addresses. A router examines the network addresses in the packets it receives and forwards data destined for the Internet to the ISP's router.

Routers bring to your connection a host of enhanced features, including the
following:

- NAT (Network Address Translation)
- DHCP (Dynamic Host Configuration Protocol)
- VPN (Virtual Private Networking)
- Firewall support

Most DSL routers also include built-in network hubs so you can create a LAN
using the DSL router, without having to purchase a hub. Unfortunately, most
of these built-in hubs support only standard 10-Mbps Ethernet and not the
newer, widely used 100-Mbps Fast Ethernet.

Routed service is typically bundled as part of a business-class DSL service
offering that includes routable IP addresses, multiple email boxes, DNS
(Domain Name System), and other services. Routed service is typically more
expensive than bridged DSL service, and routers typically cost around $450.
Many DSL routers also provide support for security and VPN.

For computers on the LAN side of the router, you can have two types of IP
address assignments: Network Address Translation (NAT) with dynamically
assigned private IP addresses or static, routable IP addresses assigned to
each computer on the LAN.

Sharing a DSL connection using NAT

NAT is an Internet standard that allows your LAN to use private IP addresses
for TCP/IP networking. These IP addresses have been set aside for use in private networks and are not routable across the Internet. Using NAT means one
Internet routable IP address is assigned to the router but the IP addresses used
by the computers on the LAN behind the router use the private IP addresses.
As a result of this translation, the computers behind the router are invisible to
the Internet. Using private IP addresses and NAT can be a cost-effective way to
connect a LAN to the Internet because you reduce the number of routable IP
addresses you need to use. NAT also acts as a simple but by no means complete firewall to protect your LAN from Internet intruders.

Most DSL routers support two NAT methods: one-to-one IP mapping and one-
to-many IP mapping. Used together, these two forms of NAT provide your
business with maximum flexibility in setting up an Internet connection. For
example, a small business with 15 computers might connect to its ISP with
just 5 IP addresses: one each for the four servers, or hosts, and a fifth IP
address to be shared, one-to-many, with the remaining computers on the

LAN. This approach offers all the standard benefits of NAT: Security from would-be intruders by hiding private IP addresses behind the NAT wall and then convenience of enabling your business to run Internet servers or hosts using routable IP addresses. It also allows the use of applications previously precluded by classic NAT, such as Web server hosting, video conferencing, and voice applications.

Chapter 7 explains NAT in more detail.

NAT is commonly used with another common DSL router feature called DHCP (Dynamic Host Configuration Protocol) to automatically assign IP addressing information to computers on the LAN whenever they boot up. This method, called *dynamic IP addressing,* saves the time and effort of manually configuring every computer on your LAN with static IP addresses.

Making your computers Internet hosts

An *Internet Protocol (IP) address* is a numeric identifier (such as 216.216.6.26) assigned to every device connected to a TCP/IP network. Each computer or other device that uses TCP/IP is distinguished from others by this unique IP address.

IP addresses are leased by ISPs in blocks of eight, sixteen, thirty-two, or more for routed service. These blocks are determined by the way the Internet subdivides itself. If you use Internet routable IP addresses as part of your routed service, each computer connected to your DSL connection is a recognizable host on the Internet. This means you can set up a Web server or any application server to be accessible by Internet users. An IP address also allows you to use an application such as video conferencing and be contacted by anyone connected to the Internet using the unique IP address assigned to the computer with a video conferencing system installed.

DSL routed service typically includes a block of eight IP addresses as part of the service package. Of these addresses, five are available for computers on the LAN. More IP addresses can be leased from the ISP for a monthly fee ranging from $25 to $100.

When you lease IP addresses from an ISP, the ISP ensures that those addresses are routable on the Internet. This means the ISP adds them into their servers so that they become part of the Internet. The ISP also assigns your custom subdomain names for each IP address, so users can connect to the computer or device using the user-friendly text name instead of the IP address.

Chapter 7 explains IP addresses and the Domain Name System in more detail.

Putting Your IP Package Together

The *Internet Protocol (IP)* is responsible for basic network connectivity on the Internet. It's at the heart of configuring your local network to make the Internet connection. The core elements of IP networking are IP addresses and DNS (Domain Name System).

An essential part of your DSL service to the Internet is setting up your IP address and DNS services. If you want computers on your LAN to become hosts that can be accessed from the Internet, you need to use routable IP addresses. You typically assign the routable IP addresses statically to specific computers on your LAN to make them reachable at the same IP address.

These hosts have full two-way communications over the Internet, which means they can support such applications as IP voice and video conferencing or operate as a Web server, email server, or any Internet-accessible server. Along with static IP addresses, you can assign secondary domain names to the hosts so that they can be accessed by using DNS.

For more detailed information on TCP/IP networking fundamentals, see Chapter 7.

IP addresses are the telephone numbers of the Internet

IP addresses are the equivalent of telephone numbers for computers connected to the Internet. Assigning a computer a specific IP address, such as 199.232.255.113, uniquely identifies that computer from all other computers connected to the Internet. If you want computers on your LAN to become hosts accessible to Internet users, you need Internet routable IP addresses — one for each computer. These IP addresses also become the basis for assigning user-friendly names to hosts using DNS.

Blocks of IP addresses are available from most ISPs for a monthly cost based on the number of IP addresses. Some ISPs include a block of IP addresses as part of the service. The IP addresses that the ISP assigns to you are available for use while you are the ISP's DSL customer. They remain the property of the ISP and return to the ISP upon termination of the service.

Routed DSL service requires three IP addresses: one for the router, one for the Ethernet connection, and one for the WAN connection. If you get a block of eight IP addresses, for example, only five are available for hosts on your LAN.

If you're using bridged DSL service, you can get IP addresses in any number because you don't use subnet routing. This means you can use a bridged DSL service with one, two, three, or more IP addresses. If you use an Ethernet router behind a DSL bridge with NAT, you need only a single static or dynamic IP address to share the DSL connection across multiple computers.

DNS service puts a name to a number

If an IP address is the equivalent of a telephone number on the Internet, a domain name is the equivalent to the name of the person or organization to which the telephone number is assigned. As people move around, their telephone numbers can change but their names do not. Your domain name can be moved to different IP addresses, but the domain name remains the same.

The process of converting domain names to machine-readable IP addresses is called name *resolution*. The Domain Name System (DNS) provides the name resolution process for the Internet using a hierarchy of DNS servers. In DNS, each host belongs to a domain. When you use both the host name and the domain name, you're using a *Fully Qualified Domain Name (FQDN)*. For example, the host name of a specific computer might be dangell, which becomes part of a domain name such as northpointcom.com to create the FQDN dangell.northpointdsl.com. The ISP will register the host names you specify for specific IP addresses.

The ISP provides name resolution service so that TCP/IP applications can use Internet domain names instead of just numeric IP addresses. The ISP operates domain name servers, which your computers and applications can access for domain name resolution. It's very important to configure name resolution because many applications depend on being able to use Internet domain names rather than IP addresses.

The DNS implements name-to-address assignments and provides a lookup to determine a computer's IP address based on its domain name. The DNS also provides reverse mapping to determine a computer's domain name from its IP address.

Make sure that your ISP provides entries in the DNS to implement reverse DNS assignments for your IP addresses and their assigned names. Many public servers on the Internet deny service to computers that do not have reverse DNS implemented or whose name-to-address mapping does not match their address-to-name mapping. Your ISP should implement default values in its name servers for your IP addresses for both name-to-address assignments and address-to-name assignments (forward and reverse DNS).

Often, failure of DNS name resolution is misinterpreted as a connectivity failure and a circuit outage, when in truth the application is simply not able to determine the IP address of the desired destination due to a problem with name resolution.

Master of your domain name

A domain name (`company.com`) allows any individual or company to maintain its own identity on the Internet. This domain name can then become part of your Internet DNS address for email addresses (`user@company.com`) and names of publicly accessible Web servers (`www.company.com`). You can use your domain name as part of your DSL service and Web and email hosting services.

Top-level domain names end in `.com` (for commercial), `.net` (for network service) or `.org` (for nonprofit organizations).

Before you can use a custom domain name for your company, you can register it with Network Solutions at `www.networksolutions.com` or with any number of other competing registration services recognized by The Internet Corporation for Assigned Names and Numbers (ICANN). You can check these registration services at `www.icann.org`.

Domain names are registered on a first-come, first-served basis, although a flurry of activity has recently surrounded the issue of trademark protection with domain name assignments.

You can register your own domain name or have your ISP do it on your behalf as part of setting up your DSL Internet service. Your ISP might charge a nominal fee, but typically the ISP registers the domain name at no charge. You still must pay a rental fee for the actual domain name. Getting a domain name to use right away costs $70 for the first two years. After that, the cost is $35 a year.

If the name is available but you haven't decided on an ISP for your DSL service, you can reserve the name. You register the domain name but it isn't assigned to a specific IP addresses. When the registration service holds a domain name until your ISP uses it for your Internet service, you pay an additional fee, over and above the $70 for renting the domain name for two years.

Grabbing a domain name and holding it until you're ready to use it could mean getting the name you want. Otherwise, if you wait until you actually sign up for your Internet service, someone might take the domain name you want.

The act of registering the domain with Network Solutions doesn't mean that the domain is functional on the Internet. The domain must be configured on the name server(s) listed on the domain registration template submitted to Network Solutions by your ISP.

If your ISP registers your domain name, Network Solutions invoices you directly for the domain registration fee. The current cost is $70.00 for registering a domain name and the first two years of use. After that, it's $35 a year. The ISP doesn't pay the domain name registration — you do.

Don't spend money creating business cards or marketing material that contains a domain name that you have not yet secured.

You can transfer your domain name

If your domain was previously hosted at another ISP or hosting service and you want to move it to a new ISP's domain name servers, a few steps are needed to ensure a smooth transition. The domain transfer process can take place independent from your circuit installation process or other changes to your service. Follow these steps:

1. **Notify your previous ISP that you will be transferring your existing domain to your new DSL ISP's name servers.**

2. **Your DSL ISP will configure the domain on its name servers and duplicate the domain's zone information, including the contents of the domain.**

3. **When the domain configuration on the ISP's name servers is verified as operational, the ISP submits a domain modification transfer request to Network Solutions.**

4. **Network Solutions emails the domain contact, which is you or someone in your business, for approval.**

   The update causes the root or master name servers on the Internet to point to the ISP's name servers as responsible for your domain information.

5. **After Network Solutions processes the update, the ISP's name servers are now authoritative sources of DNS information for your domain.**

   Your DSL ISP should notify your ISP that they should remove your domain information from their name server's configuration files because they're no longer hosting the domain. (You might need to follow up on this.)

Hosting Your Internet Email Services

Email service is an essential component of any Internet connection. When it comes to getting Internet email service, two options are typically available: hosting your email services with your DSL ISP or operating your own mail server on your network. In most cases, you use your ISP to handle your email service as part of your DSL Internet service package.

Internet email primer

Two primary electronic mail protocols are currently in use in the Internet world: POP (Post Office Protocol) mail and SMTP (Simple Mail Transfer Protocol) mail. Most ISPs support both protocols. *POP* mail client programs allow individual users to retrieve mail from their mailbox on the remote mail server. Most popular mail reading programs (such as Eudora, Netscape Messenger, and Microsoft Outlook) are POP mail client programs and can access a POP mailbox on a mail server.

SMTP is used on the Internet to deliver mail to a user's mailbox located on a mail server. An SMTP mail server handles mail for all user mailboxes at a given Internet domain, such as `anyuser@company.com`. The mailboxes for individual users are typically located on the SMTP mail server for the domain, whether it's located at an ISP or on a computer on your LAN. SMTP takes care of getting the email messages to the mail server for your domain. Individual users must then have a way to access their mailboxes on the remote mail server. POP is the most common method of accessing mailboxes.

Most mail client programs deliver outbound mail by using SMTP. POP is used for picking up mail from your mailbox; SMTP is used for sending mail to others. If you have a mail server for your domain, your client computers typically should be configured to point to your mail server as the SMTP server to use for outbound mail. Otherwise, they should be configured to send email directly to the destination by using SMTP.

In general, configuring a POP client involves simply setting up your mail program so that it knows the following:

- **The IP address or name of the POP mail server where your mailbox resides**
- **Your username on the POP mail server**
- **Your password on the POP mail server**
- **The IP address or name of an SMTP mail server for sending outbound mail**

Most POP mail client programs have many more configurable user preferences. The basics of being able to access your mailbox, however, usually require only the information listed here.

Hosting your Internet mailboxes with the ISP

For smaller organizations, hosting your Internet mailboxes with the ISP makes the most sense. Each of your users has his or her own mailbox on the ISP's mail servers that use your domain name as part of the email address, such as david@angell.com. You can typically add email boxes for a nominal fee of $2 to $10 per month.

Your users can use any POP mail client program to retrieve messages from their mailboxes. Mail from the Internet is delivered to the ISP's main server and then distributed to specific email boxes.

ISP email hosting frees you from the mail server administration and maintenance chores associated with running a mail server yourself. Users access their email directly from their own email box on the ISP server as well as from anywhere on the Internet. If you're using a dynamic IP Internet access account with no DNS support, you can use an email hosting service tied to your domain name.

Running your own email server

Installing and running your own SMTP mail server on your network allows you to manage your Internet email along with your LAN email. This management involves using an email server program, such as Sendmail, Microsoft Exchange, or MDaemon. If you elect to run your own SMTP mail server, mail for your domain is sent directly from the Internet to the mail server on your network. Connections are made directly from the mail sender on the Internet to your mail server and never touch any servers at the ISP.

Configuration and administration of the mail server is your responsibility. Typical configurations might include providing a mailbox for each user locally on the mail server or implementing mail forwarding and redirection of a user's mail to a separate internal or external mail server where the user's mailbox actually resides.

Running a mail server requires that you have a domain name of your own. The single most important requirement is that your name server be configured with an MX record for your domain. An MX, or Mail eXchanger, record tells everyone on the Internet which computer is acting as the mail server for your domain. Anytime someone sends email to one of your users, the sending program looks up the MX record for your domain to determine where to deliver the mail.

Keeping Up with Network News

Network news (also called *Usenet*) is a massive, distributed, text-based discussion group covering every topic imaginable. People who subscribe to newsgroups communicate by using a messaging system similar to email.

Two components make up Usenet. First, thousands of news servers continuously pass articles back and forth as new articles are posted and old articles are flushed out. The news servers are to a large degree independent from each other, allowing a highly distributed and robust delivery mechanism. Second, individual users read and post articles from their local news server by using news client software. News client software is typically included as part of your email or Web browsing program.

Your users may read and post Usenet news articles directly from the ISP's news servers by using the Network News Transfer Protocol (NNTP). This is the best way to go in most cases. Server-to-server Usenet news feeds are not provided with most DSL ISP service due to the sheer volume of news postings. Usenet news traffic currently averages between 10 to 15 GB per day. Delivering this amount of traffic over a DSL connection consumes all available bandwidth and makes the connection unusable for any other purpose.

Should You Host Your Own Web Server?

A DSL connection using routable IP addresses and Domain Name Service can support running a Web server on your LAN. However, evaluate the pros and cons of running your own Web server versus using a Web hosting service. Even if you use your domain name as part of your DSL Internet service package, you can use the www.yourcompany.com URL for a hosted Web server.

A Web site demands bandwidth, and if your site becomes popular, it can easily overload your DSL connection. In addition, you need to address a variety of administration and security issues in running a Web server on your LAN.

In many cases, you'll find it easier to use a Web hosting service instead of running your own Web server. Web hosting services are affordable, and you can easily manage them remotely by using a number of Web site management and content authoring tools, such as Macromedia's DreamWeaver or Microsoft's FrontPage.

Chapter 6

Shopping for Your DSL Internet Service

A DSL connection opens up the world of high-speed Internet access to the bandwidth deprived. Getting to that world, however, can be tricky business in the chaotic Wild West atmosphere of today's DSL service deployment, packaging, and pricing. Before you go shopping for DSL service, read this chapter to get a framework for navigating through the DSL shopping maze.

Your DSL Service Shopping Game Plan

The consumer adage *caveat emptor* (buyer beware) is alive and well when it comes to shopping for DSL service. Hype abounds, and just below the surface lies a host of fine-print traps waiting for the unsuspecting. What you don't know can cost you time and money. Your first order of business is to create a DSL game plan that will guide you through the process of selecting your DSL connection to the Internet.

The big three pieces of DSL service

Three key parts make up a typical DSL service offering, although most ISPs sell DSL service as a complete Internet access and connectivity package. Each of these main pieces of DSL service breaks down into sublevels of related issues. The three top-level components of the total DSL ecosystem follow:

- **The DSL circuit provider.** An ILEC or a CLEC delivers the DSL pipeline service. They operate the DSLAMs (Digital Subscriber Line Access Multiplexers) at central offices and link them to a backbone from which the ISPs then connect. Different DSL circuit providers offer different DSL flavors and speed capabilities.

- **Internet service.** The Internet service provider is often your interface to DSL service. The ISP handles the TCP/IP networking infrastructure associated with your DSL connection. ISPs buy DSL circuit services from CLECs and ILECs. What you can and can't do is defined in large part by what services the ISP offers.

- **Customer Premises Equipment (CPE).** The specific CPE you use with your DSL service plays a pivotal role in defining what is and isn't possible with your Internet service. The DSL circuit provider determines which products can be used because the CPE must work with the DSLAM equipment used by the DSL circuit provider.

DSL game plan checklist

The process of establishing a DSL connection involves evaluating interrelated components that make up your DSL service package. Here's a checklist of key questions you need to address as part of your DSL shopping game plan:

- **Who offers DSL service in your area?** This is the essential first step. If DSL service is available in your area, you need to assess what DSL flavors are being offered. In areas with competition for DSL service, you might have a choice of different types of DSL technologies, such as SDSL, ADSL, and G.lite. Each of these technologies has different capabilities, pricing, and configuration options. Most DSL service will be ADSL and G.lite from ILECs and CLECs.

- **What are your bandwidth needs?** Articulating your bandwidth needs is, at best, an educated guess, but you need a consistent benchmark for comparative shopping. Determining your bandwidth needs depends on a variety of factors. In the real world, the biggest constraint on bandwidth is cost, although doing your shopping homework can get you more speed for the same amount of money, depending on the DSL provider. The more bandwidth you want, the more it will cost. One-time start-up and CPE costs typically remain the same (or rise only marginally) as the capacity of the DSL connection increases, so the real cost increase is in the monthly DSL circuit charge. One of the nice features of many DSL flavors is the capability to upgrade your bandwidth without buying new CPE. Bandwidth is addictive and, over time, you'll want more.

✔ **What type of CPE do you need to use?** No DSL service shopping list is complete without an analysis of your CPE options. Typically, the DSL provider drives what DSL CPE can be used with the DSLAM at the CO. The biggest divide in DSL CPE is whether it supports only a single user or multiple users on a LAN. As part of your CPE decision, you also need to think about how you plan to configure your home or business for IP networking.

✔ **How good is the Internet service provider?** The ISP is in the driver's seat in defining your Internet service capabilities through DSL. The ISP providing the DSL service can put restrictions on your DSL service in its contract. For example, some ISPs set limits on the total amount of data traffic that can pass through your connection in a given month based on the speed of your connection. For example, a 160-Kbps connection might be allowed 10 gigabytes a month, but a 416-Kbps connection might be allowed up to 30 gigabytes. Some ISPs restrict you from running any type of server (including NAT for sharing the Internet connection) on your DSL connection. Because the ISP is also your interface to TCP/IP networking elements, you want to know about an ISP's technical support. Does the ISP offer technical support 24 hours a day, 7 days a week?

✔ **What kind of TCP/IP configuration do you plan to use?** You need to think about your TCP/IP networking infrastructure and how it will relate to the DSL-based Internet connection. For example, using registered IP addresses allows Internet users to access servers running on your LAN by using DNS (Domain Name System) addresses, such as www.angell.com. Likewise, you need an IP address for each host or network device that you want accessible from the Internet. For example, if you plan to use video conferencing, you want an IP address assigned to the computer running the desktop video conferencing system.

✔ **How many users do you want to connect to the DSL connection?** The number of users (computers) with which you want to share the DSL connection affects the type of DSL ISP service you want. Many ILECs package DSL Internet access service offerings that don't support multiple computers or DNS services. These solutions can be modified to support multiple users (with proxy servers or Ethernet routers), but you won't be able to provide any Internet services, such as running a Web server or video conferencing.

> ✔ **What will the complete DSL service package cost?** You need to crunch all the numbers to get an accurate estimate of what the DSL connection is really going to cost. The total price you pay for your DSL service depends on many variables, including one-time installation charges, CPE costs, and monthly service charges.

Which DSL Flavor Should I Choose?

The four DSL flavors that are the most widely deployed today are ADSL (Asymmetrical DSL), G.lite, SDSL (Symmetrical DSL), and IDSL (ISDN DSL). Within these four DSL flavors, ADSL is the most widely deployed followed by SDSL, with G.lite just getting started.

SDSL service offerings by CLECs are broken down into business-class and residential service. Business-class SDSL is typically bundled with a DSL router, static IP addresses, DNS, email, and other services tailored to business needs. Residential DSL service, which is considerably cheaper than business SDSL service, is restricted to residential installations and typically includes a DSL modem and one IP address with no DNS service.

Consumers who are typically doing more downloading of data than uploading can use the ADSL offerings with slower upstream connections. In many ADSL offerings, however, the upstream speeds are more than enough for small businesses, telecommuters, and Internet power users. In fact, ADSL can be packaged as symmetrical service, such as 384 Kbps/384 Kbps. ILECs use ADSL as the basis for both their consumer and business DSL service offerings.

The biggest danger with ADSL offerings from ILECs is that some have low upstream speeds that prevent you from using cool Internet applications such as video conferencing. Always make sure you know what the upstream speed is for any ADSL service. Some ILECs, such as Bell Atlantic, currently support only a 90 Kbps upstream speed in some of their ADSL offerings.

The only reason you would get IDSL, with its one speed of 144 Kbps/144 Kbps, is because you can't get SDSL or ADSL service. IDSL is offered by CLECs and ILECs because it has the flat-rate, always-on benefits of DSL out to a range that's double that of ADSL or SDSL. If you're using ISDN more than 20 hours per month and IDSL is available, switch over because in most cases you save money.

What's Your Need for Speed?

Bandwidth capacity planning is one of the most important and most difficult tasks you undertake in setting up a DSL connection. A number of factors come into play when trying to evaluate your bandwidth needs. Here are some basic bandwidth questions you need to answer:

- ✔ **Will you be running any servers on your DSL connections?** A big factor in determining your bandwidth requirements is whether you plan to run any servers on your DSL connection. If you plan to run a Web server, an email server, or any type of Internet server, you need to take into consideration the incoming traffic as well as your outgoing traffic.

- ✔ **Do you plan to use IP voice and video conferencing?** If used heavily, these cool (but bandwidth hungry) applications can eat up your DSL connection. Supporting multiple simultaneous users compounds the demands.

- ✔ **How do you plan to use your DSL connection?** If you will be using symmetrical applications (such as video conferencing or VPN), the upstream of the ADSL connection needs to meet the minimum qualifications. So, if you will have IP video and want a 384-Kbps/384-Kbps connection, a 1.5-Mbps/384-Kbps ADSL connection is still fine. You run into problems, however, when your symmetrical application requirements exceed the upstream of ADSL.

- ✔ **Do you plan to connect one computer or a network to your DSL line?** If you want to connect more than one computer, you're looking at a LAN-based connection to the Internet. The DSL CPE used in your connection combined with your IP addressing options define what you can do with your DSL connection. You can find DSL offerings, especially by ILECs, that offer attractive DSL speeds but use single-computer CPE with dynamic IP addressing to restrict the use of the service. You can, however, use a proxy server or an Ethernet router to share the connection.

- ✔ **Can you upgrade to a higher speed?** One of the nice features of DSL service is its bandwidth scalability. This means you can upgrade to a higher speed over time without having to start all over again. For example, SDSL service offers bandwidth options that range from 160 Kbps to 2.3 Mbps. You could start with 384 Kbps, for example, and move up in increments to 2.3 Mbps. Most SDSL bridges and routers allow you to upgrade with only minor software configuration changes. DSL service providers charge a fee for upgrading, typically in the range of $100 to $200.

Some DSL CPE bundled with DSL service packages offered by ILECs are bridges connecting only one computer to the Internet through DSL. Third-party proxy server and Ethernet router solutions let you get around this restriction to leverage your DSL connection by sharing it. See Chapter 10 for more information.

How fast does a file transfer over a DSL connection?

A *bit,* or binary digit, is the smallest unit of digital information. It takes 8 bits to make a *byte.* The designation MB stands for megabytes, or 1 million bytes. (Well, technically 1MB equals 1,048,576 bytes, but we'll use the 1,000,000 for our calculations.) Data communications is measured in bits per second (bps). Your programs and data files are measured in bytes.

To calculate the estimated time a file measured in bytes passes through a data connection measured in bits per second, you need to do a little math. Keep in mind that the result of this calculation is theorectical. In the real world, a number of variables affect the actual throughput. You can expect a reduction of speed between 10 to 30 percent from networking overhead and bottlenecks.

To estimate how much time it takes to download or upload a file using a given speed, use the following formula:

file size (MB) x 8 = total bits/DSL speed = total seconds/60 = time in minutes

For example, here's how to estimate the time it takes to transfer a 10MB file on a 784 Kbps DSL connection

10MB x 8 bits = 80,000,000 bits

80,000,000 bits / 784,000 bps (DSL speed) = 102 seconds

102 seconds / 60 seconds = 1.7 minutes

Table 6-1 provides some guidelines for bandwidth requirements of different types of Internet applications. You can use these as a rough measure of what different uses demand in terms of bandwidth ranges. Multiply the bandwidth requirements for each computer you want to connect.

Table 6-1   Bandwidth Requirements for Different Applications

Application	Minimum Bandwidth Requirement
Desktop video conferencing	384 Kbps
Distance learning	384 Kbps
Interactive video games	128 Kbps
Web surfing	56 to 128 Kbps
IP voice	56 Kbps
Teleworking	144 Kbps
Video on demand	1.5 Mbps
Web hosting	384 Kbps–6 Mbps

Choosing the Right Internet Service Provider

Depending on where you're located and the way DSL service is delivered in your area, the Internet service provider is typically the single interface to the DSL service, the CPE, and the Internet access account. Behind the ISP are CLECs and ILECs providing the circuit service and defining what CPE is used for the service. Chapter 7 goes into more detail on the IP networking considerations as part of your Internet service package.

You typically receive a single bill from the ISP for both the Internet service and the DSL circuit. This is the way most CLEC DSL services are sold. This is the easiest approach and the one that makes the most sense. In the case of ILECs, they may or may not offer ISP service as part of their DSL offerings, or they may provide the DSL connection in partnership with ISPs. Some ILECs bill the customer directly for the DSL line, separately from the Internet service provider.

ISP shopping tips

Shopping for DSL service consists of finding the right type of DSL service and the right ISP for your Internet access needs. Here are some guidelines to help you in your quest:

- Shop around to compare DSL offerings among ISPs in your area.
- Ask around to find out the ISP's reputation for service and performance
- Use the ISP's Web site, where you can often find detailed information about their DSL services.
- Understand what you need before shopping.
- Don't be afraid to ask questions.
- Get a written quote.
- Read the fine print on the terms and conditions of your DSL service. Are there usage restrictions? What are the payment terms? How long is the time commitment for the service contract?
- Ask about any promotions currently offered, such as rebates on equipment or installation fees.
- Ask whether the ISP offers toll-free 24x7 technical support.
- Find out what DSL equipment the ISP sells and supports.

Check out the ISP's Web site

One of the most important sources of information in shopping for DSL service is the Internet service provider's Web site. Unfortunately, the quantity and quality of information provided varies widely. A good Web site can save you a great deal of time.

Chapter 4 provides a detailed listing of ILEC and CLEC Web sites as a starting point for checking out DSL service. Here are some guidelines to judge a helpful Web site:

- **Does the Web site provide specifics on the ISP's DSL service offerings?** Glossy marketing copy doesn't make for good consumer information. Information packaged to educate the consumer about the service and product details is helpful.

- **Does the Web site include the costs of the ISP's DSL offerings?** Installing and using DSL service involves a variety of charges. A Web site should provide a breakdown of the costs for getting the DSL service, including a menu of optional services.

- **Does the Web site include the ISP's terms of the service?** Unfortunately, most ISPs don't post their terms and conditions on their Web sites. These documents are the fine print of your DSL service. Terms and conditions are usually provided as part of the formal quote.

- **Does the Web site provide good CPE information?** CPE is an important part of your DSL service capabilities. The Web site should provide coverage of different CPE options and their prices.

- **Does the Web site provide information on IP addresses services and Domain Name Service?** A good Web site will list your IP address options and costs, such as the additional costs for IP addresses and DNS registration. The site should also include a menu of custom services.

The devil is in the details

The terms and conditions in the DSL service contract can often make or break a DSL service deal. Restrictions on your DSL service are spelled out in this contract, so you must read it carefully to understand fully what you can and can't do with your DSL service. Here are some important things to look for in a DSL service contract:

- **What is the time commitment for your DSL service?** Many ISPs require that you make a commitment to the DSL service of one to three years. Depending on the provider, the cost per month might drop if you commit for a longer period. The trend in DSL service is toward lower prices, so it's in your interest to avoid long-term service contracts. You will rarely want a contract that lasts more than one year.

✔ **Is usage restricted?** Many ISPs include restrictions on the amount of data going through your DSL pipe per month per line. For example, there might be a 10-gigabyte limit for a lower-bandwidth DSL connection. Any data moving across your DSL line in excess of the limit costs extra, usually based on a price per megabyte. For residential service, you may find an even harsher restriction that forbids you from running any servers or using a router that supports NAT and DHCP to share the connection.

✔ **What are the payment terms?** Increasingly, ISPs are billing to credit cards to cut down on their accounts receivable overhead. Others will bill you on a monthly or annual basis, with the annual billing method offering a discount. Many ISPs also require a deposit to start the service.

Crunching the Numbers

The total price you pay for your DSL service depends on a variety of factors, but the main costs associated with DSL service are the one-time startup and CPE charges and the recurring monthly charges. Figure 6-1 provides a worksheet you can use to calculate the complete cost of your DSL service.

DSL Service Worksheet		
	ISP 1	ISP 2
One-time ISP/DSL service installation		
One-time DSL equipment cost		
Total One-Time Cost		
Monthly ISP/DSL service		
Monthly Additional IP services		
IP addresses		
Email boxes		
Other		
Total Monthly Cost		

Figure 6-1: DSL service cost worksheet.

What are the start-up charges?

Before competition heated up in DSL, DSL customers faced high start-up costs in the forms of DSL circuit activation charges from the DSL provider, Internet account set-up charges, and DSL CPE charges. Today, most DSL service start-up costs have been dramatically reduced and consolidated into a single one-time start-up charge.

Most residential DSL offerings include a free DSL modem and free installation. Business DSL offerings typically have higher start-up costs because of the added cost of using a DSL router.

Following are the charges you can expect for starting DSL service:

- ✓ **DSL circuit activation and on-site installation.** These charges include the activation of the DSL service to your premises and inside wiring. These costs can range from $99 to $200.

- ✓ **Internet access account activation.** Setting up your ISP connection may also have one-time charges for setting up the accounts you need for email, IP address services, Domain Name Service, and any custom services. Most ISPs don't charge for many of these services, so watch out if an ISP is trying to add these charges to your bill. Beyond a minimum number of IP addresses provided with the DSL service, there will be an additional charge for the first year's rental of more IP addresses. You might have to pay a registration fee if the ISP registers your domain name with InterNIC, which is a separate charge from the InterNIC fee.

- ✓ **DSL CPE.** The cost of DSL CPE depends on the type of CPE you use and the ISP's markup. DSL CPE prices range from free with a one-year commitment to $500. Some ISPs offer modem leasing or rental as part of the connection. Typical DSL routers should be around $500 to $700, and typical DSL LAN modems should go for under $300. Many residential DSL service offerings, however, include the DSL modem for $99 or free.

What are your monthly charges?

Monthly fees are usually a single, flat-rate charge based on the speed of your DSL connection. The two main components of your monthly bill for DSL service are the circuit charge from the ILEC or CLEC and the Internet service cost. The circuit charge is typically part of a single bill from the ISP or, in some cases, on a separate bill from the circuit provider.

You might also have other monthly charges for additional service beyond the basic package, such as additional email boxes and IP addresses. If you're renting your CPE, you also have a monthly rental fee.

You Ordered DSL, Now What?

To install your DSL service, both the ISP and the DSL circuit provider perform many individual tasks. At several points during the provisioning process, an ISP might provide a status report on your installation (usually through email).

The installation process for your DSL service can take from 5 to 30 business days from the date you place your order. Most DSL service installations are completed within 20 business days, and the trend is to reduce this time.

The most common cause of delay is not technical problems, but appointment and communication foul-ups between the DSL provider and the customer. In other cases, the installation might take longer due to unanticipated problems with the phone circuits available to your location. Such conditions require additional work to provision your circuit.

If you're getting ADSL or G.lite service from an ILEC or CLEC, the DSL service piggybacks on an existing POTS line, so the ILEC doesn't need to install a new telephone line. If you're using SDSL service from a CLEC, a new telephone line must be installed at your premises by the ILEC. If an unused pair available in a telephone line already going into your premises, you won't need a new line. After the new line is established by the ILEC, the CLEC sets up your DSL service and a CLEC inside wiring technician comes to your home or business to complete the DSL service installation.

With the advent of the FCC line-sharing mandate, CLECs will be able to offer ADSL service over existing telephone lines, which should reduce the time it takes to get DSL service up and running. The trend in DSL service delivery is to make the process easy enough for the DSL customer to do a self-install of the DSL CPE and service. By some estimates, 70 to 80 percent of all residential DSL installations will be performed by the customer.

Part II
Making DSL Come Alive with Internet Access

The 5th Wave   By Rich Tennant

WE'RE HERE TO
SERVICE YOUR
LOCAL-AREA-NETWORK.

In this part . . .

A DSL connection comes alive with Internet access. This part covers the essential elements you need to grasp to get the most out of your DSL Internet service. You get help in understanding the fundamentals of TCP/IP (Transmission Control Protocol/Internet Protocol) as a foundation for choosing the right IP network configuration for your business or home.

Next, you explore your DSL CPE (Customer Premises Equipment) options and what the differences between a DSL modem and router mean to you. After mastering TCP/IP and DSL CPE essentials, you move into the realm of Internet security to understand the risks of an always-on Internet connection. I guide you through a variety of affordable solutions for protecting computers and valuable data from hackers. Finally, I show you how to share a DSL Internet connection when the ISP supplies only one IP address.

Chapter 7

All Things TCP/IP

. . . .

. . . .

TCP/IP is the universal language for computer communications on the Internet. The type of DSL CPE (modem or router) and corresponding TCP/IP networking configuration you use define your Internet service. Understanding your TCP/IP options empowers you to make good choices in getting the most out of your DSL Internet service. This chapter explains the fundamentals of TCP/IP networking as a backdrop for making the right decisions about your DSL CPE and Internet access service.

TCP/IP Delivers the Internet to You

TCP/IP, or Transmission Control Protocol/Internet Protocol, is the lingua franca of the Internet. A *protocol* is a set of rules used to allow interoperability among different systems. TCP/IP is actually a combination of two key protocols as well as a collection of other supporting protocols. *Transmission Control Protocol (TCP)* is the transmission layer of the protocol and serves to ensure reliable, verifiable data exchange between hosts on the network. (A *host* is any computer or other device connected to a network.) TCP breaks data into packets, wrapping each with the information needed to route it to its destination, and then reassembles the packets at the receiving end of the communications link.

TCP puts in the packet a *header* that provides the information needed to get the data to its destination and then passes the packet to the *Internet Protocol (IP),* which is the networking layer of TCP/IP. IP inserts its own header in the packet and then moves the data from point A to point B.

The five layers of TCP/IP

TCP/IP takes a layered protocol approach to networking, which means each protocol is independent of the others but all protocols work together to enable TCP/IP networking. TCP/IP loosely follows the Open Systems Interconnection (OSI) model, which defines how computer-networking devices should communicate with each other. The OSI model defines the framework for implementing protocols in seven layers: application, presentation, session, transport, network, data link, and physical.

The TCP/IP protocol consists of five layers — application, transport, Internet, data link, and physical — that perform the functions of the OSI model's seven layers. The OSI model's session, presentation, and applications layers are combined in the TCP/IP's application layer. The five layers of TCP/IP are described in Table 7-1.

Table 7-1	The Five Layers of TCP/IP
Layer	*What It Is*
Application (layer 5)	The layer where you do your work, such as sending email or requesting a Web document from a Web server. At this layer, you're working with protocols that form the basis of TCP/IP applications, such as HTTP for the Web and FTP for file transfers.
Transport (layer 4)	The layer that makes sure your packets have no errors and that all packets arrived and are in the correct order. This layer transports data through TCP and passes it to the Internet layer.
Internet (layer 3)	The layer where the IP (Internet Protocol) fits into the equation. This layer gets packets from the data link layer and sends them to the correct network address.
Data link (layer 2)	The layer that splits your data into packets to be sent across the connection medium. It interfaces with the physical layer, which is the hardware and network medium.
Physical (layer 1)	The pure hardware layer, including NICs, cabling, and any other piece of hardware used on the network.

When information passes from computer to computer through TCP/IP networking, control of the data passes from one layer to the next, starting at the application layer and proceeding through to the physical layer. The information then proceeds to the bottom layer of the next system and up the hierarchy in that second system.

Why layers matter

Why is it useful to understand the OSI and TCP/IP layers? You often see references to different DSL CPE working at different layers of the OSI and TCP/IP network models. This difference defines the two leading DSL CPE devices: bridges and routers. A DSL modem (a bridge) works at the OSI and TCP/IP layers 1 and 2, but DSL routers go up to layer 4. The higher the layer at which a DSL CPE works, the more sophisticated its capabilities. Chapter 8 goes into more detail on DSL CPE options.

You also see the layers of the OSI model commonly referenced in Internet applications. For example, different VPN (virtual private networking) technologies use different layers. VPN is the technology used to create a private, secure link (called a *tunnel*) through a public network such as the Internet. A layer 2 (data link layer) VPN solution means that any network protocol such as IP, Microsoft's NetBEUI, and others can be used as the basis for communicating across the Internet. A layer 3 (network or internet layer) VPN solution is tied to a specific networking protocol, which is typically IP.

The life of packets

TCP/IP is based on a technology known as packet-based networking. In a *packet network,* data travels across a network in independent and variable-sized units that can be routed over different network paths to reach the destination.

The life of a packet (also called a *datagram*) begins when an application, such as a Web browser, creates it. As each packet travels down the sending computer's layers, it picks up additional control and formatting information in a header to ensure its delivery to the destination computer. Each router on the network that encounters a packet examines the header to determine whether the packet is intended for its local network. If not, the packet is passed on in a direction closer to the ultimate destination. When the packet reaches the destination computer, the header is read and stripped as the packet moves up through each layer.

An important benefit of the packet-switched network is that packets in a message don't have to travel to the destination along the same route. Instead, packets can travel many different routes; they all end up at the same destination, where they're reassembled in the order originally intended. This independent routing of packets over a network allows data to be transmitted even if parts of the network are disrupted.

Keeping packets moving with routing

The process of getting your data from point A to point B through the TCP/IP network is called routing. Routers forward all packets they receive that aren't part of their known IP address universe to another predetermined router. This forwarding, or *routing,* continues until the packet reaches its destination. The entire path to the destination is known as the *route.* The number of routers in a transmission path between two hosts on the Internet is referred to as the number of *hops.*

A router allows data to be routed to different networks based on packet address and protocol information associated with the data. Routers read the data passing through them and decide where the data is sent. This decision-making functionality, called *filtering,* allows or disallows certain source IP ranges, or protocols, from either entering or passing. With filtering, a router can help protect your network from unwanted intrusion and prevent selected local network traffic from leaving your local network. Filtering is not perfect, however, and can be hacked. Chapter 9 explains Internet security issues.

Any port in a storm

On a TCP/IP network, data travels from a port on the sending computer to a port on the receiving computer. A *port* is a unique address that identifies the application associated with the data. TCP uses the port to figure out which application is sending or receiving data. Table 7-2 lists common Internet application port addresses.

Table 7-2	Common TCP/IP Port Numbers
Internet Service	*Port*
World Wide Web	80
FTP	21
Email (SMTP)	25
Network News	144

A computer on a TCP/IP network "listens" for data traffic associated with a specific application through the port identifier. For example, a Web browser or server monitors port 80, and an FTP client or server monitors port 21.

Typically, port numbers above 255 are reserved for private use of the local machine, and numbers below 255 are defined as defaults for a variety of universal TCP/IP applications and services. TCP/IP doesn't see the specific application. It sees only the numbers — the Internet address of the host that provides the service and the port number through which the application intends to communicate.

Creating Identity with IP Addresses

Within any networking protocol, there must be a way to identify individual computers or other networked devices. TCP/IP is no exception and includes an addressing scheme that pervades the Internet and intranets. Think of Internet Protocol (IP) addresses as the unique telephone numbers for specific computers or other network devices on any TCP/IP network.

An *IP address* is a software-based numeric identifier assigned to each machine on an IP network. Each computer or other network device that uses TCP/IP is distinguished from others on the network by this unique IP address. IPv4 IP addresses are 32-bit addresses and are usually represented as decimal values between 0 and 255. An IP address is organized into four groups of 8-bit numbers separated by periods, or dots, such as 199.232.255.113.

IP addresses are very difficult for humans to remember, so easier-to-remember *domain names* are mapped to each IP address. In that way, we humans can refer to a specific computer or device by its domain name rather than its IP address. The Domain Name System (DNS), which is explained later (in the "What's In a Name Anyway?" section), provides the friendly text interface to IP addresses.

Are you dynamic or static?

An essential piece of your DSL service is the type of IP address configuration: a dynamic IP address Internet service or a static IP address Internet service. The key distinction here is that using a *dynamic IP* type of Internet access account means that your PC or LAN is invisible to the Internet. This configuration is targeted at the consumer who doesn't plan to run any type of Internet server.

Dynamic IP addressing complicates your use of Internet applications that require an IP address, including video conferencing, VPN, games, and other applications. Because your IP address changes, anyone trying to connect to you through the IP address will always need to know the new IP address. My advice: If you have a choice, avoid dynamic IP address Internet service.

Static IP addresses are recognized and routable on the Internet. Static IP addressing is typically used by businesses and power users to enable them to run Internet servers and such applications as Net voice and video conferencing. Static IP address Internet access accounts cost more because you must lease IP addresses from the ISP.

Both dynamic and static IP addresses are Internet-routable IP addresses, unless you're using NAT with private IP addresses for computers on your LAN. A computer or device using an Internet-routable IP address is a host on the Internet, which means the computer or device is accessible to Internet users.

IP addresses have class and no class

Traditionally, IP addresses were assigned to networks using three classifications based on size: Class A, Class B, and Class C. This breakdown of IP addresses was simply a way of allocating addresses among the different networks that access the Internet:

- ✔ **Class A networks** are the El Grande of IP networks. Only 126 Class A addresses are possible in the world, and each Class A network can have in excess of 16 million computers in its individual networks.

- ✔ **Class B networks** can have up to approximately 65,000 workstations on the network. Approximately 16,000 Class B networks are in the world.

- ✔ **Class C networks** can have up to 254 workstations on the network. Several million Class C networks are possible.

Until 1994, Class C addresses were the smallest block of IP addresses that could be assigned. Many smaller companies that need IP addresses don't need a Class C network with 256 IP addresses. Likewise, some companies need more than a Class C but less than a Class B, and so on. In response to the limitations of A, B, and C Classes of IP addresses, Classless Internet Domain Routing (CIDR, which is pronounced "cider") was implemented. CIDR networks are described as *slash x* networks, where *x* represents the number of bits in the IP address range. Table 7-3 lists a sampling of slash x network configurations that support a specific number of IP addresses.

Table 7-3	Common Slash x Network Configurations	
Network Type	*Subnet*	*Number of IP Addresses*
Slash 27	255.255.255.224	32
Slash 28	255.255.255.240	16
Slash 29	255.255.255.248	8
Slash 30	255.255.255.252	4

Ipv6: The next generation

The current version of IP addressing is IPv4 (for version 4). Over at least the next ten years, IPv6, a new generation of IP, will be phased in. Aiding the gradual transition is the fact that IPv4 and IPv6 can coexist. One of the main reasons behind IPv6 is that the Internet is approaching the 4 billion addresses limitation of IPv4. IPv6, however, offers a virtually unlimited number of new IP addresses. Here's a sample IPv6 address:

```
EFDC:BA62:7654:3201:EFDC:BA72:
   7654:3210
```

IPv6 retains most of IPv4's characteristics, but many important things change for the better. IPv6 provides a wide range of improvements for IP networking, including the following:

✓ **Big changes in the way you get IP addresses.** In the IPv4 environment, you have to contact your ISP and get a new IP address. After you get the address, you then configure the computer with the IP address information and update a router. This process can take hours or days, depending on your ISP and its technical support. IPv6 uses a new collection of features called autodiscovery, autoconfiguration, and autoregistration. Together, they provide easier management of a network with no manual intervention. By using a system of queries and Plug-and-Play, the network can detect and automatically assign an address. Autoregistration is the way IPv6 handles dynamic adding, which is the process of updating a computer's host name and address information in DNS.

✓ **Improved security enhancements.** Security services such as packet authentication, integrity, and confidentiality are part of the design of IPv6. Because these security services are built right into IPv6, they are available to all TCP/IP protocols, not just specific ones such as SSL, PPTP, and S-HHTP. This means your organization can easily be made more secure. The new IPSec protocol, a component of IPv6, adds an additional level of security for IPv6 and creates a secure, TCP/IP-level, point-to-point connection. IPv6 also offers security to applications that currently lack built-in security and adds security to applications that already have security features.

✓ **Better living through multimedia.** IPv6 provides new capabilities for high-quality, streaming, and multimedia communications, such as real-time audio and video.

Subnet subdivision

TCP/IP networks can procreate by subdividing into smaller networks called subnets. A *subnet* is a collection of computers that can communicate with each other without the need for routing. Subnets are created using the host portion of an IP address to create something called a *subnet address,* or subnet mask. This IP address allows the workstation to identify the network of which it is a part. When you use a DSL router with static IP addressing, you create a subnet. If you use a DSL bridge, your LAN is part of the ISP's subnet.

Unlisted numbers: Private IP addresses

With the proliferation of TCP/IP as the networking protocol of choice for LANs, organizations wanted placeholder IP addresses that they could use for their private networks. The IETF (Internet Engineering Task Force), the group responsible for implementing and maintaining Internet standards, set aside a Class A, B, and C series of IP addresses that can be used exclusively for intranets. These *private IP addresses* can't be assigned on the Internet and will not route through the Internet. Therefore, any organization of any size can use these IP addresses for their intranets. Those building their own intranet don't have to lease IP addresses from an Internet service provider to set up a TCP/IP network.

The address ranges reserved as private IP addresses are as follows:

10.0.0.0–10.255.255.255 for Class A networks

172.16.0.0–172.31.255.255 for Class B networks

192.168.0.0–192.168.255.255 for Class C networks

Using these addresses on your local network makes them invisible to the Internet. These private IP addresses hide behind the single registered IP address used by the router with NAT activated.

Listed numbers: Routable IP addresses

By using Internet-routable IP addresses, you can do a lot more with your DSL connection. IP addresses linked to specific hosts and domain names enable Internet users to access a host computer running as a Web server (or any TCP/IP application server) using the user-friendly text identifier, such as www.angell.com. This makes you a provider of Internet services for all Internet users or a private workgroup.

A configuration using a block of registered IP addresses on a LAN might break down as follows:

✔ One IP address goes to a Windows NT/2000 Server to be used by Microsoft Internet Information Server (IIS) running as a Web server. With a single IP address assigned to the Web server, you can run multiple virtual Web servers off the single IP address.

✔ One IP address is assigned to the Windows NT/2000 Server to be used by Microsoft Exchange Server or another email server for email.

✔ Multiple IP addresses are assigned to client computers (hosts) to enable two-way Internet access.

You get routable IP addresses from the ISP. Most DSL Internet service includes a basic package of IP addresses tailored to the type of service. Depending on the service you're using and the ISP, you might be able get additional IP addresses for an additional fee. Your ISP will also register your domain name and enter subdomain names into their DNS server, if DNS service is offered as part of the DSL Internet service package. After you get your domain name, you can have the ISP register the subdomains for all your IP addresses on their DNS server.

Using static and dynamic IP addressing

As you've read, you can have private IP addresses and routable IP addresses. Both of these IP address types can be assigned to computers or other network devices on a static or dynamic basis.

A *static IP address* is a fixed IP address assigned to a specific computer or other device on your network. The IP address remains associated with that computer or device so that it can be accessed from the Internet. Think of an IP address as a telephone number that connects to a specific residence. For every computer or device you want available to Internet users, you need an assigned static IP address. For a computer running Microsoft Windows, you enter the IP address as part of configuring Microsoft TCP/IP.

Before you can configure the computer on your LAN for static IP addressing, you need the following information from your ISP:

- An IP address for each workstation
- A subnet mask IP address for your network
- The IP address assigned to your DSL router or the ISP's router (if you're using a DSL modem), which is the gateway IP address
- A host name and a domain name for the registered IP addresses
- DNS server IP addresses

Most DSL routers include a built-in DHCP server to enable dynamic IP addressing on your LAN. The DHCP server automatically manages the assignment of IP addresses, subnet masks, and default gateway addresses to computers as they sign on. The server then manages the IP address table, making sure that only one address is assigned to each active workstation. From a user's perspective, these negotiation and assignment procedures are transparent.

In most cases, you use DHCP with NAT (Network Address Translation) and private IP addresses on your local network. DHCP is a typical feature in DSL routers. All Microsoft Windows clients (Windows 95, 98, NT Workstation, and 2000) and the Mac can act as DHCP clients.

It's All in the Translation

Network Address Translation (NAT) is an Internet standard that allows your local network to use private IP addresses, which are not recognized on the Internet. The IP address used for the router as a gateway is provided by the ISP as part of the DSL service. Figure 7-1 shows how NAT works on a DSL connection.

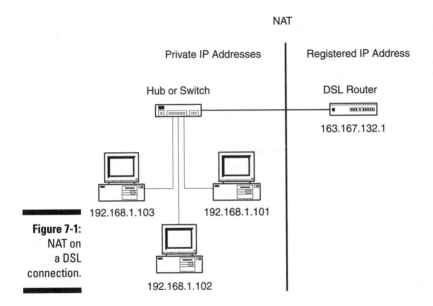

Figure 7-1:
NAT on
a DSL
connection.

The computers behind the NAT can access the Internet through the router, but Internet users can't access the computers behind the router. You can allow traffic from the Internet to pass through NAT, however, by mapping ports to specific private IP addresses. For example, you can configure NAT so that it lets Web browser requests from the Internet pass through port 80 to connect to a computer running a Web server on your local network.

Using private IP addresses with NAT, smaller organizations can realize significant cost savings because only a single-user Internet access account is required for connecting an entire LAN to the Internet. NAT also provides increased security because the IP addresses used on the intranet are unrecognizable on the Internet.

An inherent problem with NAT is it doesn't support Internet applications that use dynamic TCP ports, such as video conferencing, games, and streaming audio and video. Fortunately, most routers today provide workarounds that allow these and other applications to pass around NAT.

Even if you're using a DSL modem (bridge) to connect to the Internet, you can use NAT by adding an Ethernet router, which sits behind the DSL modem. The Ethernet router includes the same functionality as the router built into the CPE device. You can also use other Internet-connection-sharing solutions, such as proxy servers. Chapter 10 explains your options for sharing a DSL connection using a single IP address and NAT, if you're using a DSL modem.

More than one way to NAT

Using NAT doesn't restrict you from using one or more computers on your local network as hosts on the Internet. Most DSL routers support two NAT methods: one-to-one IP mapping and one-to-many IP mapping. These two forms of NAT used together provide you with maximum flexibility in setting up an Internet connection. For example, a small business with fifteen computers might connect to its ISP with just five IP addresses: one each for the four servers or hosts and a fifth IP address to be shared, one-to-many, with the remaining computers on the LAN. This approach offers all the standard benefits of NAT — the security by hiding private IP addresses behind the NAT wall from would-be intruders and the convenience of enabling your business to run Internet servers or hosts using routable IP addresses.

One-to-one IP mapping is also embodied in a router feature called *DMZ* (Demilitarized Zone). A DMZ port allows one or more PCs on a LAN behind NAT to be a host on the Internet to provide unrestricted two-way communications for such things as running a Web server or video conferencing. DMZ enables the router to translate an Internet-routable IP address (such as 216.216.6.8) to the private IP address assigned to the PC on the local network (such as 192.168.1.101).

NAT and DHCP go together

NAT is commonly used with another common DSL router feature called DHCP (Dynamic Host Configuration Protocol) to automatically assign the IP addressing information to computers on the LAN whenever they boot up. This method, called dynamic IP addressing, saves the time and effort of manually configuring every computer on your LAN with static IP addresses. DHCP can be used to randomly assign IP addresses whenever anyone on the local network boots up or the DHCP server can always assign a specific IP address to the same computer each time it boots up.

What's in a Name Anyway?

If an IP address is the equivalent of a telephone number on the Internet, a *domain name* is equivalent to the name of the person or organization that the telephone number is assigned to. As people move around, their telephone numbers change but their names do not. Domain names provide the friendly text interface to IP addresses.

The process of converting domain names to machine-readable IP addresses is called *name resolution*. During the resolution process, a computer called a DNS (Domain Name System) server translates the domain name into an IP address.

When you provide an address for most Internet operations, such as pointing your Web browser to a Web site or sending email, you can use either the IP address or the domain name method. Most organizations use domain names as their form of addressing because they are easier to read and understand. Your domain name can be moved to different IP addresses, but the domain name always remains the same.

Host names, domains, and subdomains

Domain names are based on a hierarchical structure. Each period, or dot, within the domain name identifies another sublevel of the overall organization through which the message must pass to arrive at its destination. The order of levels in a domain always proceeds left to right from the most specific to the most general, such as `david.sales.angell.com`.

Each host belongs to a domain, and each domain is classified further into domain levels. The format of a host name with a domain name is

> HostName.DomainName

The Domain Name System provides a centralized online database for resolving domain names to their corresponding IP addresses.

Domain names can be further subdivided into subdomains. These are arbitrary names assigned by a network administrator to further differentiate a domain name. The format of a host name with both subdomain and domain names is

> HostName.SubdomainName.DomainName

Both the domain and subdomain names further describe a particular computer.

Top-level domain names

At the end of all Internet addresses is a three-letter domain level such as
.com or .edu, which is referred to as a top-level domain. These *top-level
domains* provide an indication of the organization that owns the address, and
they always appear at the end of the domain name. The purpose of the top-
level domain is to provide another level of distinction for a full domain
address. Seven organizational domains are available, as shown in Table 7-4.

Table 7-4	Organizational Domains
Organizational Domain	*Entity*
.com	For-profit commercial organizations
.edu	Educational institutions
.gov	Nonmilitary government organizations
.int	International (NATO) institutions
.mil	Military installations
.net	Network resources
.org	Nonprofit groups

There are also geographic domains, such as .au for Australia, .uk for United
Kingdom, and .jp for Japan. Geographic domains indicate the country in
which the name originates. In almost all instances, the geographic domains are
based on the two-letter country codes specified by the International Standards
Organization (ISO). You typically find many domain names that include the
.com in front of the two-letter country code, such as www.angell.com.uk.

A new crop of top-level domains

Like so many other areas of the Internet, the pool of available domain names
ending with the traditional three-letter, top-level domain name is dwindling
rapidly. As a result, a new crop of top-level domain names could become
available for use on the Internet.

The Internet Corporation for Assigned Names and Numbers (ICANN), which
is comprised of Internet standard-setting bodies and legal and communications experts, is working on new top-level domain names. ICANN also accred-
its companies to act as competing registration firms to handle Internet
registrations.

The domain name game

Your domain name is registered with Network Solutions or another ICANN accredited registrar, which you can check out at www.icann.org. They handle the registration services for .com, .net, .org, .edu, and .gov top-level domain names.

The cost to register a domain name varies from $25 to $35 a year. Names are registered on a first-come, first-serve basis. Registering a domain name implies no legal ownership of the name.

A top-level domain name establishes your business name as defined in the Domain Name System. From the top-level domain, you can register subdomain names directly from your ISP. These are the addresses that define host machines and servers, such as www.angell.com or david.angell.com.

Domain names are not case sensitive, so it doesn't matter whether letters are uppercase or lowercase. No spaces are allowed in a domain name, but you can use the underscore (_) to indicate a space. You can use a combination of the letters *A* through *Z*, the digits 0 through 9, and the hyphen. You can't use the period (.), at sign (@), percent sign (%), or exclamation point (!) as part of your domain name because DNS servers and other network systems use these characters to construct email addresses.

After you have decided on your domain name, you should check to see whether it's available. Because of the immense popularity of the Internet, finding a domain name that isn't already in use is becoming harder. You should do your own checking of domain names before you decide on a domain name and submit it your ISP.

The act of registering the domain does not mean that it's functional on the Internet. The domain must be configured on the name server(s) listed on the domain registration template for your ISP.

When you register your domain name through your ISP, make sure that the registered party is yourself, not your ISP. If you change ISPs, you want to make sure that you own the name and can transfer it to your new ISP.

IP Configuration Recipes

At this point, you might be totally confused about IP addresses and domain names and what they have to do with your DSL connection. To put this information in a tangible form that you can relate to, here are some real-world IP configuration recipes that you might use when setting up your DSL service.

Security issues are also involved in making your local network accessible to the Internet, but for now focus on IP address and domain name issues. Security issues are explained in Chapter 9.

A block of Internet-routable IP addresses

In the first example, or recipe, you have a block of registered IP addresses and their associated domain names, which means the gateway device and the computers on the local network are accessible from the Internet.

Blocks of IP addresses are available from most ISPs for a monthly cost based on the number of IP addresses. Some ISPs include a block of IP addresses as part of the service. The IP addresses assigned to you by the ISP are available for use while you are the ISP's DSL customer. They remain the property of the ISP and return to the ISP if you terminate the service. When you buy IP addresses from an ISP, the ISP ensures that those addresses are routable on the Internet and should include custom subdomain names, such as `www.angell.com`, `david.angell.com`, and so on.

Bridged, or routed, DSL service requires three IP addresses just for the IP server. One IP address is used for the router, one for the Ethernet connection, and one for the WAN connection. This means that if you get a block of eight IP addresses, only five are available for hosts on your LAN.

For routed DSL service, you can get IP addresses as a block of eight, sixteen, or thirty-two IP addresses. These IP addresses typically cost about $25 a month per block of eight addresses. These blocks of IP addresses are determined by the inherent characteristics of subnet routing. If you're using bridge DSL service, you can get IP addresses in any number because you don't use any subnet routing.

Suppose you have a small network of five computers. One of these computers is a server running Windows NT Server. You also plan to use the computer as a Web server using Microsoft Internet Information Server (IIS). In addition, you want to assign IP addresses to four computers that can be running Windows 95, Windows 98, or Windows NT Workstation as their operating system. As part of your DSL service, you plan to use a DSL LAN modem and use email boxes hosted by your ISP.

One IP address goes to a Windows NT Server to be used by Internet Information Server running as a Web server. With a single IP address assigned to the Web server, you can run multiple virtual Web servers off the single IP address.

Table 7-5 lists examples of IP addresses and domain names for this scenario. Figure 7-2 shows how this IP address configuration would look on a small network.

Table 7-5	Static IP Address and Domain Name Recipe for a Small Network	
IP Address	*Subdomain*	*What It Is*
209.67.232.2	www.angell.com	The IP assigned to your Web server. When users on the Internet enter the URL www.angell.com in their browser, they connect to your Web server running on the Windows NT Server computer.
209.67.232.3	catie.angell.com	The IP address assigned to a host computer named catie on the LAN using the angell.com domain.
209.67.232.4	david.angell.com	The IP address assigned to a host computer named david on the LAN using the angell.com domain.
209.67.232.5	joanne.angell.com	The IP address assigned to a host computer named joanne on the LAN using the angell.com domain.
209.67.232.6	starr.angell.com	The IP address assigned to a host computer named starr on the LAN using the angell.com domain.

NAT and private IP addresses

This section describes a scenario for a small office that is interested in only jacking up the speed of Internet access to a small network of five users. The business is currently using several POTS modem accounts and wants to consolidate them into a single DSL connection. This business will use a DSL router that includes NAT and optional DHCP features. The IP address used for the router as a gateway is provided by the ISP as part of the DSL service.

These machines can access the Internet through the router, but Internet users can't access the computers behind the router. The router, however, can use TCP/IP ports to allow certain types of TCP/IP application traffic, such as port 80 for Web browsers connecting to a Web server on your local network.

Figure 7-3 shows a LAN-to-DSL connection using a router with NAT activated. Notice that the computers behind the NAT use private IP addresses that aren't recognized on the Internet.

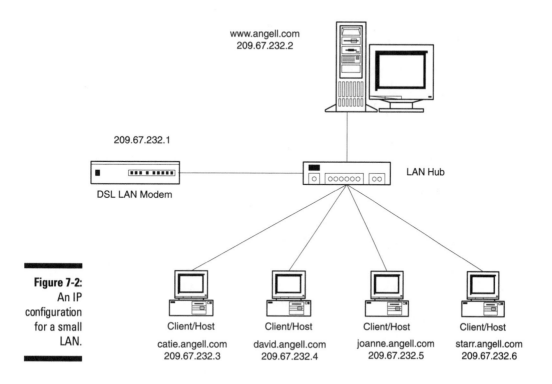

www.angell.com
209.67.232.2

209.67.232.1

DSL LAN Modem

LAN Hub

Figure 7-2:
An IP
configuration
for a small
LAN.

Client/Host

catie.angell.com
209.67.232.3

Client/Host

david.angell.com
209.67.232.4

Client/Host

joanne.angell.com
209.67.232.5

Client/Host

starr.angell.com
209.67.232.6

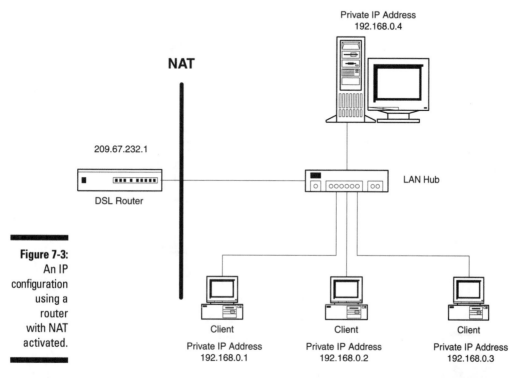

Private IP Address
192.168.0.4

NAT

209.67.232.1

DSL Router

LAN Hub

Figure 7-3:
An IP
configuration
using a
router
with NAT
activated.

Client

Private IP Address
192.168.0.1

Client

Private IP Address
192.168.0.2

Client

Private IP Address
192.168.0.3

These private IP addresses are not assigned on the Internet and will not route through the Internet. Therefore, using these addresses on your local network makes them invisible to the Internet. These private IP addresses hide behind the single registered IP address used by the router with NAT activated. For example, with five computers on your local network, you can use the private IP address range of 192.168.0.1 through 192.168.0.5 and a subnet mask of 255.255.255.0.

With your private IP addresses defined, you can choose to assign a specific IP address to each computer on your LAN or use the DHCP feature of the DSL router to automatically assign IP addresses from the pool to each computer on an as-needed basis.

Fun with a NAT combo

If your DSL router supports the two NAT methods — one-to-one IP mapping and one-to-many IP mapping — you can create a hybrid IP configuration (Figure 7-4) that allows you to use both routable and private IP addresses together.

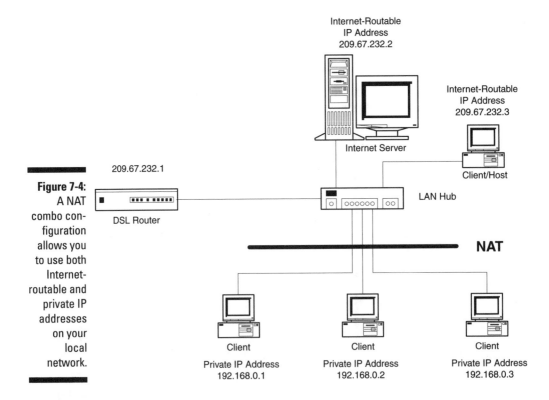

Figure 7-4: A NAT combo configuration allows you to use both Internet-routable and private IP addresses on your local network.

Suppose you have a small network of eight computers. You want two of these computers to be Internet hosts with Internet-routable IP addresses. One of these computers is a server running Windows NT/2000 Server and Microsoft Internet Information Server for a Web server. The other computer is connected to a video conferencing system to allow people in your office to video conference clients. The remaining five computers don't need to be hosts on the Internet; people in your office use these computers for routine Internet activities such as Web surfing, email, and file transfers. Therefore, these computers don't need routable IP addresses. Figure 7-4 shows how this combo Internet-routable and private IP configuration works.

PPP over DSL

The Point-to-Point Protocol (PPP) is a TCP/IP protocol that enables dial-up connectivity to the Internet. This is the protocol you use whenever you connect to the Internet through a dial-up modem or an ISDN modem. In Microsoft Windows 95, 98, NT, and 2000, PPP-based connections are handled through Dial-Up Networking (DUN).

A growing number of ADSL and G.lite DSL consumer offerings are using PPP over ATM (PPPoA) or PPP over Ethernet (PPPoE). These technologies create a dial-up connection over an always-on DSL line. When you use PPP over DSL to connect to the Internet, you're connecting to the ISP's network. (You make an instant connection through the DSL network using a virtual circuit to the ISP's network.) You're not dialing up the ISP the way you do with an analog modem. As with a dial-up connection, you log on to the ISP's network using a username and a password.

PPP over DSL allows ISPs to leverage their existing dial-up networking infrastructure to handle DSL Internet connections. For DSL subscribers, PPPoA and PPPoE have serious drawbacks. DSL providers and ISPs can easily oversubscribe PPPoA connections like they do for dial-up connections, which means if a lot of people are trying to connect to the Internet, you might not connect. PPPoA and PPPoE also allow ISPs to shut off your connection at any specified time intervals they choose. In addition, installing a PCI or USB DSL modem that uses PPPoA can be klugey because it installs as a network adapter but uses dial-up networking to make a connection.

Chapter 8

Your DSL Equipment Field Guide

DSL service ultimately connects your computer or local area network to the Internet. What stands between you and this connection is the Customer Premises Equipment (CPE). The type of DSL CPE you use as part of your DSL service plays an integral role in defining your Internet service. This chapter gives you a solid grounding in the differences in your DSL CPE options.

What Does Your DSL CPE Do?

DSL CPE provides two essential functions for your DSL Internet connection. The first is converting the digital content from the computer to a form of digital signaling used for the data communications link. This function is generically called a CSU/DSU (Channel Service Unit/Data Service Unit). The CSU terminates the digital data communications line. The DSU converts the digital signals coming from a computer networking device (such as a bridge or a router) into a digital signal understood by the data communications link.

The second function of your DSL CPE is handing IP (Internet Protocol) networking to enable your computer or LAN to connect to the Internet through your ISP.

DSL CPE comes in two flavors: bridges and routers. Bridges and routers are internetworking devices used to allow separate networks to communicate with each other. DSL providers and CPE vendors commonly use the term *DSL modem*. This is more a metaphor than a descriptive term to describe the modem-like functionality of a DSL modem in making the connection to the Internet through DSL.

A new DSL CPE product will be hitting the market to support Voice over DSL (VoDSL). This product incorporates an Integrated Access Device (IAD) that enables voice communications to be carried over a DSL line as data packets until they reach the CO, where the data is converted and passed over to the PSTN network. IADs will most likely be integrated into DSL CPE. VoDSL and IADs are covered in more detail in Chapter 18.

How DSL CPE connects to your computer (s)

DSL CPE connects to the DSL line at one end and to your computer or LAN at the other. DSL CPE can connect to your computer using one of the three following interfaces:

- ✔ **Ethernet.** Ethernet forms the basis of the most popular form of computer networking for PCs and Macs. You can connect Ethernet DSL CPE (modems or routers) to a single computer or a network. Ethernet is the most commonly used interface for DSL CPE because it can easily support the speeds of DSL (up to 100 Mbps) and enables the DSL connection to be shared across a network. Ethernet DSL CPE can be used with PCs and Macs.

- ✔ **USB (Universal Serial Bus).** USB supports data speeds of up to 12 Mbps between your computer and peripherals. USB also allows you to easily connect computer peripherals, including a DSL modem, to a computer without cracking the case to insert an adapter card. USB ports have been standard on most PCs for the last few years; Microsoft Windows 98, Windows 2000, and the latest Macs support USB. A USB modem is an external device that connects to a USB port on your PC and is typically used for consumer (single-computer) DSL offerings, although there are ways to share the connection.

- ✔ **PCI (Peripheral Component Interconnect).** PCI is a standard for adapter cards that insert into a slot inside your PC. You can't use PCI adapter modems on a Mac. Installing a PCI DSL modem isn't as easy as installing an external USB modem. As is the case with USB modems, PCI DSL modems are typically bundled with consumer (single-computer) DSL offerings, although there are ways to share the connection.

If you're planning to share a DSL connection, in most cases you're better off going with Ethernet-based DSL CPE. Even with consumer DSL offerings that include an Ethernet DSL modem and a single IP address, you can choose from several ways to share the connection. You can share USB or PCI DSL modems, but you don't have as many options for sharing as you do with an Ethernet router, and you still need to use a network to share the USB or PCI modem with other computers.

Chipsets, firmware, and DSL CPE

At the heart of any DSL device is its brain, which is commonly referred to as a *chipset*. A DSL device's functionality is determined by what is embedded in its chipset. What makes different DSL CPE work with different DSLAMs is the interoperability of their chipsets. As standards such as G.lite and others become formalized, they're embedded into chipsets to make the DSL CPE comply with the standard.

The trend in hardware is to allow software upgrades to chipsets; this is accomplished by including memory that doesn't get erased when the power is turned off. This memory, typically flash memory, contains instructions called *firmware*. A chipset with flash memory enables the firmware. This allows you to download software from the vendor's site to perform a do-it-yourself firmware upgrade. With DSL service and standards still in flux, many DSL CPE vendors include firmware in their devices to allow for changes in capabilities.

DSL CPE Shopping Tips

The DSL CPE — and the DSL circuit — are sold through the ISP. As DSL matures, you can expect to see DSL CPE sold in the retail channel as well as built into computers from PC vendors, such as Dell and Compaq. Like DSL service pricing, costs of DSL equipment vary from one ISP to another. Most ISPs, however, try to keep CPE prices low because they represent one of the biggest startup costs in getting DSL service. Pricing for your DSL CPE typically ranges between free (for DSL modems) to $600 for routers. Many ISPs offer free or reduced price DSL modems as part of an installation promotion for a one-year (or more) commitment.

The types and brands of DSL CPE available from a particular ISP are determined by the DSLAM used by the DSL provider whose service the ISP is offering customers. ISPs often support and recommend one or two brands; usually it's best to stick with the brands they know.

The bottom line is that your CPE choices are restricted by which ISP and DSL provider you use for your DSL service. The good news is that a wide variety of DSL CPE are available, and new vendors are entering the DSL CPE market with new products. In addition, the costs of DSL CPE is on a downward trend as economies of scale take hold.

Another important feature in most DSL CPE today is that you can download firmware upgrades for them from the vendor's Web site. This means the DSL equipment you buy today has a good chance of not becoming obsolete with the emergence of standards, unless you change your DSL service.

Most business DSL offerings are bundled with DSL routers, which are more expensive. However, routers include more features than bridges for managing the Internet connection, including support for VPN (virtual private networking), firewalls, and NAT (Network Address Translation). Routed Internet service usually includes static IP addresses to make every PC on the LAN a host on the Internet.

DSL modems are considerably less expensive and are commonly used with lower cost consumer and SOHO (small office/home office) DSL offerings. Typically, these DSL modem offerings include only one static IP address to restrict the number of computers you can connect to the service. A growing number of DSL providers are offering consumer DSL service kits that include the DSL CPE through retail outlets such as Best Buy, CompUSA, and Staples. You buy the kit, sign up for the service, and install the CPE yourself.

Here are guidelines for helping you choose the right DSL CPE for your needs:

- Always ask what type of DSL CPE the ISP is offering as part of their Internet service package.

- Check out the DSL CPE vendor sites to get detailed information on the specifics of their DSL CPE products. See "DSL CPE Vendor Guide" later in this chapter.

- Avoid DSL modems that recognize data traffic from only one NIC. For more information on these DSL modems, see the sidebar titled "MAC attack: Single-computer modems."

- If the Ethernet DSL modem or router includes a built-in hub, does it support both standard Ethernet (10 Mbps) and Fast Ethernet (100 Mbps)? Also check to see whether the hub has an uplink port for connecting another hub to it to expand your LAN.

- Does the Ethernet DSL modem include a separate port for connecting to a single computer and another port for connecting to a hub, so you don't have to use a crossover cable?

✔ Make sure the DSL CPE can be upgraded by installing new firmware. New Internet and DSL standards, as well as new applications, are a fact of life and you want to preserve your CPE investment.

✔ Before choosing a DSL router, ask the vendor about the security and VPN solutions they offer. Depending on the DSL router vendor, these features might be added to the router at no charge or sold as separate upgrade options. Also ask the router vendor about getting onsite service for setting up the firewall and VPN upgrades.

DSL Modem or Router?

DSL CPE comes in two forms: bridges (DSL modems) and routers. Bridges and routers are internetworking devices that enable separate networks to communicate with each other. A bridge combines one or more networks into a single seamless network. A router forwards data traffic between separate networks. DSL bridges come in Ethernet, USB, and PCI flavors; DSL routers are Ethernet devices.

Bridge over DSL waters

A *bridge* is a simple device that has to decide only whether the packet is intended for the local network or the remote network. An Ethernet bridge knows the hardware addresses of the NICs of computers connected to the network. The bridge reads the destination hardware address in a packet and decides whether the packet is going to a host on the same network or across the bridge to a different network. USB and PCI modems are also bridges.

Bridged Internet service is ideal when you're connecting less than 10 computers. ISPs offer bridged access to the Internet to allow them to manage the routing overhead and complexity on their network. Because bridges are simple devices, they're easily added to the ISP's network, without any configuration of the device on your end. And unlike routers, bridges don't need to be configured with specific IP address information. In a bridged service configuration, the ISP supplies you with a single routable IP address or multiple IP addresses. DSL bridges typically cost a few hundred dollars less than DSL routers. However, bridges don't include the enhanced capabilities of a router.

Although DSL bridges are cheaper and easier to install than DSL routers, they expose your computer or LAN to Internet hackers. Bridges lack data-filtering capabilities, so there is no basic checking of incoming packets. Fortunately, you can use a number of affordable security solutions to add protection to your DSL Internet connection. For more on security solutions, see Chapter 9.

Going the DSL router route

A router keeps your LAN as a separate network from your ISP's network. What routers bring to your connection are a host of enhanced features, including NAT (Network Address Translation), DHCP (Dynamic Host Configuration Protocol), support for VPN (Virtual Private Networking) and security (firewall) services.

A router is a more sophisticated gateway device than a bridge. A router allows data to be routed to different networks based not on hardware addresses (as in a bridge) but on packet address and protocol information. Routers work at the Internet, data link, and physical layers of the TCP/IP network model. This decision-making functionality, called *filtering,* not only enables a router to protect your network from unwanted intrusion but also prevents selected local network traffic from leaving your LAN through the router. This is a powerful feature for managing incoming and outgoing data for your site.

Two inherent features in most DSL routers are NAT and DHCP. The NAT feature lets you use private (that is, unregistered) IP addresses for your local network workstations. These private IP addresses aren't recognized on the Internet, so computers behind the NAT router remain invisible to Internet users. Using these private IP addresses costs you nothing and also makes your local network inaccessible to outsiders by keeping local traffic separate from Internet data traffic. A DSL router that supports NAT also offers basic security for your local network because NAT blocks incoming access to local computers from the Internet.

DHCP allows the router to assign IP addresses automatically to computers on a LAN as they start up. The DHCP server built into the router dynamically assigns IP addresses from a pool of IP addresses, which can be private IP addresses or public IP addresses (IP addresses recognized on the Internet). For more information on NAT and DHCP, see Chapter 7.

DSL routers are typically bundled as part of a business-class DSL service package with multiple IP addresses. A router gives you more flexibility in your IP configurations by supporting two NAT methods: one-to-one IP mapping and one-to-many IP mapping. With these features, a small business with fifteen computers, for example, might connect to the Internet using just five IP addresses. Four of these addresses can be assigned to computers to make them accessible from the Internet to support such applications as Web and email servers and video conferencing. The fifth IP address is shared, one-to-many, with the remaining computers on the LAN. This approach offers all the standard benefits of NAT: the security by hiding the private IP addresses behind the NAT wall from would-be intruders and the convenience of enabling you to run Internet servers or hosts using routable IP addresses.

Routers are considerably more expensive than modems and must be configured. Although initial router configuration might be performed for you as
part of your DSL service installation, you might to reconfigure the router as
you make changes your Internet service. Many DSL routers are difficult to
configure and might require a professional installer.

If you want a router only so that you can use NAT to share a DSL connection,
getting a DSL modem and adding an Ethernet router behind it might be a less
expensive option than going with a DSL router. A new generation of easy-to-
use, affordable Ethernet routers are available. These Ethernet routers sit
behind a DSL modem to let you use the single IP address to support multiple
computers connected to a LAN. Chapter 10 explains these Ethernet router
solutions in more detail.

Plugging into USB or PCI DSL CPE

Universal Serial Bus (USB) and PCI (Peripheral Component Interconnect) DSL
CPE are designed primarily for consumer DSL service. USB and PCI DSL CPE
are typically less expensive than Ethernet DSL CPE, which makes them attractive to consumers.

You install a USB or PCI DSL modem in your Microsoft Windows computer,
not as a modem but as a network adapter. From your PC's perspective, a USB
or PCI DSL modem appears as a network interface card (NIC). You configure
the TCP/IP settings for your PCI or USB modem using the Network properties
dialog box in Windows 95, 98, or NT and the Local Area Connection
Properties dialog box in Windows 2000.

Because USB and PCI DSL modems are treated as network adapters in your
computer, you can share your DSL connection. If your home or office has
more than one PC that you want to connect to the DSL line, sharing a USB or
PCI connection requires installing a LAN. After the network is installed, you
can use Internet-connection-sharing or proxy server software. The computer
connected to the USB or PCI DSL modem acts as the gateway for all the computers connected to the LAN.

If you use a USB or PCI DSL modem, you can't use an Ethernet router to share
the connection. You must use either a proxy server or Internet-connection-
sharing software. For more information on sharing your Internet connection
using a proxy server, see Chapter 10.

Say goodbye to an old UART

UART (Universal Asynchronous Receiver/ Transceiver) was the computer interface used for the serial port. The serial port has a limit of 115-Kbps capacity. DSL modem speeds simply overwhelm the UART's limited throughput, which is why DSL CPE uses Ethernet, USB, or PCI to connect to your computer.

Snapping in a USB DSL modem

USB DSL modems are external modems. A USB cable connects the modem to a USB port on your computer or to a USB hub. A USB hub connects to a PC to support more USB devices. Figure 8-1 shows a USB modem from Efficient Networks, and Figure 8-2 shows a USB modem from Alcatel.

Make sure your computer is running Windows 98 SE (Second Edition) or Windows 2000 before installing a USB DSL modem. If you're using Windows 98, upgrade to Windows 98 SE to avoid USB problems inherent in Windows 98. USB is not supported in Windows 95 or Windows NT.

Figure 8-1: A USB DSL modem from Efficient Networks.

Photo courtesy Efficient Networks

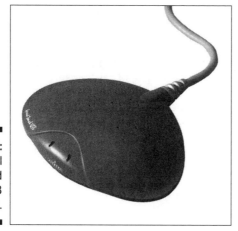

Figure 8-2:
The Alcatel
ADSL Speed
Touch USB
modem.

Photo courtesy Alcatel

USB modems are usually more expensive than PCI DSL modems but they are also easier to install (because they're external Plug-and-Play devices). In addition, you can move a USB modem from one computer to another.

USB close up

USB ports have been standard on most PCs for the last few years. Microsoft Windows 98 SE, Windows 2000, and the latest Macs support USB. USB is rapidly emerging as the preferred interface for connecting a wide range computer peripherals.

The beauty of using USB devices is that they're easier to install. You simply snap the connector into your computer or USB hub; no case cracking required. You can connect up to 127 USP peripherals to a single computer. However, most computers have only one or two USB ports, so you need to use USB hubs to expand the number of available USB ports.

To connect a high-speed device to your PC or hub, the USP cable length can be no more than 5 meters (a little under 16 feet 5 inches). For low-speed devices, the limit is 3 meters (9 feet 10 inches). You can extend the reach by connecting a USB hub to your PC and then connecting a device to the USB hub.

USB specification 1.1 supports data speeds up to 12 Mbps, which is more than enough to handle a DSL modem. USB specification 2.0, which will be supported soon, will support speeds up to 480 Mbps.

Playing your PCI modem card

As with USB connections, DSL PCI adapter card modems are designed as a low-cost, single-computer DSL solutions. DSL PCI modem cards are typically the least expensive type of DSL modem. These cards insert into a PCI (Peripheral Component Interconnect) slot. This means you need to crack open the computer case to install the card. Some DSL providers, however, might install the card for you.

The DSL line connects directly to the card. Most of these cards are Plug-and-Play compatible to make installation easy. Figure 8-3 shows the Xpeed ADSL PCI adapter.

Figure 8-3:
The Xpeed
ADSL PCI
adapter.

Photo courtesy Xpeed

PCI/USB DSL CPE and PPP over DSL

PPP (Point-to-Point Protocol) is the TCP/IP protocol that enables dial-up connectivity to the Internet. (For more on this topic, see Chapter 7.) Some DSL providers use PPP over DSL as a way to deliver DSL service. The two leading forms of PPP over DSL are

- PPP over ATM
- PPP over Ethernet (PPPoE)

If you're using a DSL service that uses PPP over DSL, you need to use a USB or PCI modem. The modem installs as a network adapter in your computer but you make a connection as if you're using a dial-up modem.

How can this be? DSL is an always-on connection! Although DSL is an always-on service, a PPP connection to the Internet is not always on. You need to use Dial-Up Networking (DUN) in Microsoft Windows 95, 98, NT, or 2000 or Remote Access on the Mac.

In Windows, after you install the USB or PCI modem, which is recognized as a NIC in Windows, you use DUN to make the Internet connection.

You initiate an Internet connection using a PCI or USB modem in a manner similar to using a dial-up modem with a username and a password, except you don't use a telephone number. Instead, you use a Virtual Circuit Identifier number.

If you're using a USB or PCI modem with PPP over DSL, the vendor typically includes software that creates the DUN profile for you as part of the DSL modem installation software. After the DUN profile is created, you might need to configure the TCP/IP settings for the connection. And to do that, see the chapter for your computer's operating system in Part III.

Going with Ethernet DSL CPE

The Ethernet interface offers the most flexibility for your DSL connection and is the most widely used computer interface. Because Ethernet forms the basis of the most popular form of networking, you can easily integrate a DSL connection into a LAN. You can also use Ethernet DSL CPE to connect a single computer with an Ethernet adapter installed.

USB and PCI DSL CPE include only modems, but Ethernet DSL CPE includes both modems and routers. Figure 8-4 shows an SDSL modem from Xpeed. Figure 8-5 shows the Efficient Networks SpeedStream 5660 ADSL router. Figure 8-6 shows the Cisco 827 Business ADSL router.

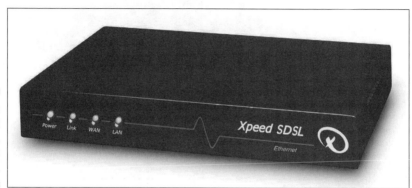

Figure 8-4:
An Ethernet
SDSL
modem from
Xpeed.

Photo courtesy Xpeed

Figure 8-5:
Efficient
Networks
SpeedStream
5660 ADSL
router.

Photo courtesy Efficient Networks

Figure 8-6:
Cisco 827
Business
ADSL
router.

Photo courtesy Cisco Systems

MAC attack: Single-computer modems

MAC, or Media Access Control, is the protocol that operates at the data link layer of the OSI protocol stack. It controls access from the NIC to the network media (cabling). Each NIC has its unique MAC address, which is also called a *hardware address.* (An example MAC address is 00-80-AE-00-00-01.)

Bridges accept TCP/IP packets from NICs. A LAN bridge learns all the MAC addresses of all NICs on the local network to determine what is and isn't local data traffic. A single-computer bridge used for a DSL connection recognizes traffic from only the single MAC address of the NIC it's connected to.

Most Ethernet routers and some modems include a built-in network hub so you can build your own small LAN using DSL CPE. Every computer you connect to your Ethernet DSL CPE must have an Ethernet adapter or NIC (network interface card) installed. And every computer on the network connects to the hub using 10BaseT cable. Today, creating your own network is easy and inexpensive, and new network technologies make networking even better, such as wireless or Phoneline networking. Chapter 11 shows you how to build your own LAN.

Surveying DSL CPE Vendors

DSL CPE vendors fall into two camps. The first group are traditional telecommunication equipment companies, such as Alcatel, that supply the telephone companies with DSLAMs as well as CPE. Most computer users aren't familiar with these companies because they've sold their products primarily to telephone companies. The second group of vendors comes from the computer networking and internetworking industry. Some might be familiar: 3Com, Cisco, Efficient Networks, and Netopia.

Table 8-1 lists DSL CPE vendors, their Web sites, and a brief description of their CPE line. Check out their Web sites for more detailed coverage of their DSL CPE offerings. The DSL CPE business is competitive and new products and features are popping up all the time.

Table 8-1		DSL CPE Vendors
Vendor	*Web Site*	*Description*
Alcatel	www.usa.alcatel.com	Large French telecommunications company that supplies DSLAMs to most ILECs. As a result, they supply CPE to most ILECs as well. Alcatel offers a line of DSL CPE, including USB and PCI modems, Ethernet bridges, and routers. Alcatel's DSLAMs are ADSL, so their CPE is all ADSL. As of this book's writing, Alcatel doesn't support G.lite.
Efficient Networks/ FlowPoint	www.efficient.com	Sells a complete line of DSL CPE to both ILECs and CLECs. Their SpeedStream line includes ADSL and G.lite USB and PCI modems, Ethernet SDSL modems, and ADSL and SDSL routers. In 1999, Efficient Networks acquired FlowPoint, one of the leading SDSL router vendors.
Netopia	www.netopia.com	Offers, primarily to CLECs, a complete line of DSL CPE. There products include an SDSL router that supports the CopperMountain DSLAM, an SDSL router that supports the Nokia DSLAM, an IDSL router, and an SDSL modem.
3Com	www.3com.com	The leading networking products (NICs, hubs, etc.) and analog modem company has entered the DSL CPE space with a growing line of modem and router products. The 3Com line includes an ADSL Ethernet modem, ADSL and G.lite PCI and USB cards, an ADSL router, an SDSL modem and router, and an IDSL modem. Works with DSLAMs from CopperMountain and Nokia.
Cisco	www.cisco.com	Owns the internetworking router business but as of this writing has a limited presence in the DSL CPE market. The Cisco product line includes a more limited number of DSL CPE products. These products include Ethernet ADSL and SDSL modems and routers in the 600 series, Cisco 800 series IDSL routers, and Cisco 1400 series ADSL routers.

Vendor	Web Site	Description
Xpeed	www.xpeed.com	Provides a line of inexpensive DSL modems, including an IDSL PCI adapter, an SDSL PCI adapter, an SDSL USB modem, an SDSL Ethernet modem, and ADSL and G.lite PCI modems. Works with the Copper Mountain DSLAM.
Cayman Systems	www.cayman.com	Sells a line of DSL CPE, including ADSL and G.lite modems, ADSL and G.lite routers, an SDSL router, and ADSL modem router. Sells primarily to ILECs.
WebRamp	www.webramp.com	Includes a line of DSL CPE including an IDSL router, an SDSL router, and an ADSL router.
Westell	www.westell.com	Telecommunications company that offers a few ADSL modems and HDSL equipment. One of the DSL modem providers to Bell Atlantic.
Paradyne	www.paradyne.com	Makes a line of MVL DSL CPE and DSLAMs sold to CLECs.
ZyXEL	www.zyxel.com	Offers a small line of IDSL, ADSL, and SDSL routers.

What's a Residential Gateway?

A residential gateway takes a Swiss-Army-knife approach towards integrating the separate functions of networking, internetworking, voice, and broadband connectivity into a single box. A typical residential gateway combines the functions of a router, a network hub (Ethernet, Phoneline, or wireless), and DSL CPE for sharing an Internet connection.

As its name implies, the residential gateway box acts as a central, one-box solution for linking an in-home network to a DSL Internet connection. A residential gateway is based on a modular design that allows you to mix and match different technologies for a complete data communication solution. You can add advanced telephony features such as PBX functionality to your voice communications and the capability to route audio, video, and networked games.

The leader in residential gateways is 2Wire (www.2wire.com). 2Wire offers a line of residential gateways called the HomePortal; Figure 8-7 shows the 2Wire HomePortal. Other residential gateway vendors include Telocity (www.telocity.com), ShareGate (www.sharegate.com), and Boca Research (www.bocaresearch.com).

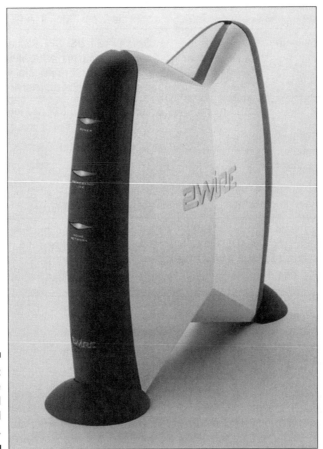

Figure 8-7:
2Wire
HomePortal
residential
gateway.

Photo courtesy 2Wire

Chapter 9

To Serve and Protect: Internet Security

The Internet is a double-edged sword. On one side, its open TCP/IP protocols and networking environment facilitate information sharing, improve connectivity, and provide greater access to the underlying network. On the other side, the same open protocols and networking systems that make the Internet popular also make security a big issue.

You need to think about security as part of your DSL Internet service package. Potential attacks by Internet hackers can destroy you data or expose confidential information. The good news is that affordable firewall solutions are available for small businesses, telecommuters, and consumers.

This chapter helps you navigate through the Internet security maze and guides you through a variety of solutions.

Locking the Door: Internet Security 101

Security is the soft white underbelly of DSL or any broadband service. Being at least a little paranoid about the security of your computer or LAN from Internet intruders is healthy. Somewhere between the hyper-paranoia pitches of firewall vendors and denials of any security risks is a rational middle ground for protecting your computer or LAN.

With always-on connections and static IP addresses, you can face a variety of security threats. These threats include the destruction or theft of data from unauthorized access and denial of service (DoS) attacks, which disable a network so that users can't access network resources.

You might not be as big a target as these big companies, but even a single attack can be devastating. Your data can be wiped out or your credit card information stolen. And in most cases, you won't be aware of the destruction or stolen information until it's too late.

To make security problems worse, the tools available to hackers today are easier than ever to use. Hackers no longer need to be skilled in attacking networks because user-friendly hacker helper programs are readily available on hacker sites.

The mantra of Internet security

The mantra of Internet security is minimizing unsolicited inbound connections. This means you must control what types of inbound connections you allow.

You need to allow inbound connections for essential Internet services, such as incoming email or Web server responses to Web browser requests from your LAN users. Beyond these basic access requirements, the more inbound connections you allow, the more you increase the potential risks.

Internet security options never offer an ironclad solution because security is a constantly evolving game of cat-and-mouse. As hackers develop new ways to break into networks, new safeguards are developed. Adding security to your network is a trade-off between unfettered, two-way access to the Internet and severe restrictions on what you can and can't do with your connection. The other big trade-offs are cost versus risk and performance versus risk.

Common TCP/IP attacks

Connecting your local network to the Internet with TCP/IP can open the doors to a variety of attacks from outsiders to gain access to your private resources or to raise havoc to your local network or Internet connection. The three common forms of TCP/IP-based security attacks follow:

✔ **Denial of service (DoS) attack.** The goal of a denial of service attack is not to steal information but to disable a device or a network so users no longer have access to network resources. Flooding an emailbox is a common use of this threat. Any type of service for which users rely on timely access is vulnerable to this type of attack.

✔ **Network packet sniffing.** Programs called *packet sniffers* capture packets from the local area network and display them in a readable manner. The source and destination users of this information probably never even know that the information has been tapped.

✔ **IP spoofing.** An *IP spoof* is when an attacker masquerades as a trusted IP address. In an IP spoof attack, the IP address of a particular machine on the Internet or within an intranet is controlled and managed entirely by the intruder, not by the owner or administrator of that machine.

VPN is a close relative of security

VPN (virtual private networking) is related to security in that it creates a secure tunnel between two points (your home computer and the company network) for your private data to travel over the Internet. Because of the relationship between VPN and security, many security solutions include VPN capabilities as an add-on to firewall functions. See Chapter 17 for more information on virtual private networking.

DSL CPE and security

Your DSL line connects to your computer or LAN using a DSL modem (bridge) or a router. DSL modem Internet connections using static IP addresses are the most vulnerable to Internet attackers. A bridge passes all packets and application information from the Internet to your network without screening incoming data. The good news is that having a DSL modem connection in no way prevents you from adding protection using a variety of security solutions.

Enhancing security with Ipv6 and IPSec

Security services (such as packet authentication, integrity, and confidentiality) are part of the design of IPv6, which is the next-generation IP networking protocol. These capabilities ensure that packets are from the correct sender, haven't been altered in transit, and can't be seen by hackers.

Because security services are built into IPv6, they'll be available to all TCP/IP protocols. Keep in mind, however, that the full implementation of IPv6 could take years. The new IPSec (IP Security) protocol is a component of IPv6 and is available for IPv4. The IPSec protocol adds an additional level of security and creates a secure, TCP/IP point-to-point connection. IPv6 offers security to applications that currently lack built-in security and adds security to applications that already have security features.

Routers include basic packet filtering and NAT (Network Address Translation), which adds some security to your DSL Internet connection. A router's packet-filtering capabilities can be customized to increase security, but doing so isn't really an option for the nontechnical user. Most DSL routers include only a minimal amount of security filtering.

DSL routers are often billed as including a firewall, but in many cases what the vendor or ISP is claiming as a firewall is NAT functionality. Although NAT does offer protection by hiding computers behind the router using private IP addresses, NAT isn't a complete firewall solution. However, the DSL router vendor might offer a built-in firewall solution or, for an additional charge, you can add a firewall.

Practicing Safe Internet Access

Although adding security to your DSL connection is important, you can also perform routine tasks to increase your security. These tasks can be as simple as turning off your DSL CPE or shutting down your computer when you're not using it for extended periods of time. The most common threat is a virus sent through email, which you can nip in the bud by running up-to-date antivirus software. Read on to review the ways you can practice safe Internet access.

Windows weaknesses

Microsoft Windows, Internet Explorer, and Outlook are prime targets of hackers. And Microsoft has made it worse by integrating browser functionality and Internet communications into the operating system.

Internet Explorer (IE) is a popular target because of its ActiveX enhancements and Java support. Because IE can execute ActiveX programs contained in Web pages, hackers have figured out different ways to download to your PC and then execute programs that can do all kinds of damage.

What kinds of damage can a hacker do by finding security holes in Internet Explorer, Outlook, and Windows? Plenty. A dedicated hacker or unscrupulous Web site operator can disable the SSL (Secure Socket Layer) security feature of your Web browser, which makes it easier to grab your credit card data from a transaction; steal your password; run destructive ActiveX programs on your system; read files on your PC or intranet; crash your system; copy your files; and more.

Microsoft regularly issues security patches to try and keep up with hackers, but *you* need to keep up with these updates. For Windows 98, run the Windows Update feature and download and install all the security patches. These are also available at `www.microsoft.com/windows98/downloads/corporate.asp`. Windows 98 SE includes some of these patches, so do likewise. Windows 95 users should also get every security patch they can at `www.microsoft.com/windows95/downloads`.

If you're running Windows 95, you're extremely vulnerable to a denial-of-service attack called *packet flooding,* where someone floods your PC with useless data and shuts down your connection and your computer.

Don't lose your cookies

Cookies are small files that Web sites save to a user's PC to identity the computer accessing the Web site. When a site knows who you are, it can present you with customized content to match your preferences. Cookies can be placed on a PC without the user's knowledge to monitor online activities and gather private information.

You don't always need to be always-on

An always-on connection doesn't always have to be on. Leaving your DSL modem on and connected is fine if your PC is turned off because a hacker can't extract much from a DSL modem. If you use your computer a lot and don't want to be always online, however, simply turn off your DSL modem.

Before turning off your DSL modem, check with your ISP/DSL provider, just to be sure. In some situations, turning off your DSL modem (as opposed to disconnecting it from your PC) triggers a line down problem with your provider. Turning the DSL modem back on might change the IP address the provider uses; older modems that store configuration data in volatile memory might require a service visit to be properly reconfigured.

Be vigilant about viruses

The number one security threat, no matter how you connect to the Internet, is a virus hiding in email attachments or in files you download from the Internet. A *computer virus* is a program that spreads copies of itself throughout a computer or network by attaching those copies to host files (usually program files, known as executable files) or email messages. Viruses typically perform a number of additional disruptive actions.

If you use Outlook Express or Outlook, turn off the mail preview feature. This feature, which automatically opens email for you, can also launch a hidden virus. With this feature off, you can scan message headers and delete any suspicious email (especially those from unknown correspondents sending you attachments). Disabling mail preview also prevents you from unintentionally activating hidden viruses and scripts.

Viruses are everywhere on the Internet, and you need to protect your PC or LAN from the havoc they can create. Although proxy servers and firewalls deal with the dangers of TCP/IP networking, they're not effective deterrents against viruses.

The best first-level defense is antivirus software, such as Norton AntiVirus or McAfee VirusScan. Make sure you have the latest version of the program and update the virus definitions weekly or sooner, if you hear any news alerts about new threats. Good antivirus software performs the following functions:

- **Prevents a virus from infecting your system in the first place.** The software recognizes hundreds of viruses by maintaining a database of virus fingerprints, or signatures. It can scan new files that a user downloads from an external source or places into a diskette drive and then raise an alert before the user copies or executes the program.

- **Provides an early warning system.** Antivirus software can detect and alert you to a virus that slips through your early warning system. You can then repair and disinfect the affected file before the virus spreads and forces your entire LAN to shut down.

- **Recovers from a virus attack.** Good antivirus software can clean, repair, and disinfect the computer or at least tell you when to throw in the towel and restore from a clean backup.

Viruses can make your life miserable. When it comes to combating viruses, an ounce of prevention is worth a pound of cure.

Fighting Fire with Firewalls

A *firewall* is a barrier to attacks from Internet intruders. They're you first line of defense against a variety of security threats. The primary purpose of a firewall is to control access from the Internet into your network.

The key differences between firewalls are the amount and quality of information used to make decisions. The more information collected, the less likely it is that an intruder will get through the firewall. Firewalls differ also in their architecture and features. The three types available today follow:

✔ **Packet filter firewalls** provide basic network access control based on protocol information in the IP packet. This information is compared to a collection of filtering rules when the IP packet arrives at the firewall. These rules specify the conditions under which packets should be passed through or denied access. Packet filter firewalls built into most DSL and Ethernet routers offer only minimal protection from Internet intruders.

✔ **Application firewalls,** commonly called proxy servers or session-level firewalls, go beyond basic packet filtering. A proxy server accepts or rejects data traffic based on an entire set of IP packets associated with an entire application session to the same IP address. For example, Web browsers generate multiple data packets as they request files from various parts of the Internet. These individual data packets are all part of the larger session. The proxy server ensures that the client connection terminates at the firewall and that a new connection is initiated to the internal protected network or vice-versa. Proxy servers provide better security than packet filter firewalls, but they also slow your connection speeds and are more difficult to set up because of their elaborate security checking.

✔ **Stateful packet inspection firewalls** are based on packet-filtering techniques but implement additional security features. Instead of just checking addresses in incoming packet headers, the stateful packet inspection firewall intercepts packets until it has enough to make a determination as to the state of the attempted connection. Packets that are cleared are forwarded to the internal network, allowing direct contact between internal and external systems. Typically, stateful packet inspection firewalls are faster than application firewalls but don't provide the same security level.

Packet filter firewalls

A *packet filter firewall* offers basic network access control based on protocol information in the IP packet. This information is compared to a collection of filtering rules when the IP packet arrives at the firewall. These rules specify the conditions under which packets should be passed through or denied access. Most packet filter firewalls are incorporated into DSL routers because packet filtering to restrict access is a logical extension of the router's functionality. As far as Internet users are concerned, the only accessible machine on the inner network is the specified host machine. Packet filter firewalls are generally considered less secure than proxy servers.

At a minimum, all firewalls use the information found in the IP packet to make decisions, as described in Table 9-1. These basic components are the protocol, the destination IP address, the destination IP port, the source IP address, and the source IP port. Some firewalls accept or reject packets by using the destination and source IP addresses and IP port information; other firewalls use the protocol in the packet. The firewall can track packets to determine who may be attempting to access the network and issue alarms to help detect suspicious activity as it occurs.

Table 9-1	Components Used by a Firewall in an Internet Environment
Component	*What It Is*
Protocol	Transmission Control Protocol (TCP) or User Datagram Protocol (UDP)
Destination IP address	Identifies the location of the computer receiving the data transmission
Destination IP port	Identifies the application on the computer that will receive the data transmission
Source IP address	Identifies the location of the computer initiating the data transmission
Source IP port	Identifies the application on the computer initiating the data transmission

Acting on your behalf: Proxy servers

Proxy servers go beyond the basic packet-filtering mechanism. They accept or reject data traffic based on an entire set of IP packets that are part of an entire session to the same address. For example, Web browsers such as Internet Explorer generate multiple data packets as they request files from various parts of the Internet. These individual data packets are all part of the larger session. Session-aware proxy servers provide better security than just plain packet-filtering firewalls.

Proxy servers are session-aware firewalls shuttle information from the original connection to the second connection. They sit between your LAN and the DSL modem (bridge) to the Internet. In essence, the proxy server masquerades as the destination computer to the network client and as the network client to the destination computer.

Proxy servers provide these key benefits:

- ✔ **Sharing your DSL connection.** A proxy server allows multiple computers on a LAN to share Internet access from a single IP address. This means that you can share a DSL connection offered as a single-user solution for your entire LAN.

- ✔ **Securing your LAN from outside intruders.** A proxy server provides relatively tight security. Attacks based on IP spoofing can't reach the local network. Computers on the LAN access the Internet indirectly through the proxy server.

- ✔ **Better utilization of your DSL connection.** Some proxy servers also include Web caching. A Web-caching proxy server cruises the Web and examines pages that your LAN users have visited and that have been cached on the server. If a page has been modified, the proxy server stores a new version on a local drive. It can also use certain guidelines to hit links on that page to pull down related pages. This saves time because users don't have to access the Internet for frequently used resources.

Stateful packet inspection firewalls

Stateful packet inspection is a technology similar to that used in enterprise-level firewall products. With stateful packet inspection, a firewall makes security decisions based on the origination of Internet sessions. It allows data to come through from the Internet only if it's part of a session initiated by one of the users on the secure LAN and blocks all communications initiated from the Internet. Stateful packet inspection has the added benefit of being easy to manage, making it ideal for organizations that don't have the technical resources for a packet filter or proxy firewall.

A central cache within the stateful packet inspection firewall keeps track of the state information associated with all network connections. All traffic passing through the firewall is analyzed against the state of these connections to determine whether it will be allowed to pass through or rejected. Stateful packet inspection uses the state information embedded in TCP packets.

Security Solutions Field Guide

Traditionally, Internet security products were designed and priced for the needs of the enterprise computing market. These firewall products were targeted at larger organizations because they had high-speed Internet connections and needed the security to protect their networks. These firewall products are simply too complicated and too expensive for smaller businesses, teleworkers, and consumers.

With the advent of affordable high-speed connections through DSL, a growing number of companies are now offering a variety of affordable hardware and software security solutions. These security solutions are in the following categories:

- **Routers.** Most DSL and Ethernet routers come with built-in filtering capabilities as well as NAT (Network Address Translation) to add basic security. These devices are built with routing in mind, however, so their security protections are limited unless you add a firewall. The NAT feature of DSL routers offers some protection to a LAN by using private IP addresses, but it's not a complete firewall solution.

- **Proxy servers.** The software proxy server runs on a computer with two NIC cards installed. One NIC sends Internet data traffic to the DSL CPE, and the other NIC connects to the local network. The proxy server works as the gateway between traffic from the local network and traffic going to the outside network. As with routers, proxy servers let you share the DSL connection using a single IP address.

- **Security appliances.** Also called firewalls in a box, security appliances provide a high level of security by combining sophisticated stateful packet inspection and basic routing functionality with an easy-to-use interface. A growing number of vendors are offering these firewall appliances for smaller businesses and SOHOs, at a cost of $299 to $995, depending on the device's capabilities.

- **Personal firewalls.** These are inexpensive, user-friendly software firewalls that install on each computer connected to the Internet. They provide basic security at the PC by monitoring the data traffic coming into the PC from the Internet.

Routers and security

DSL and Ethernet routers offer basic security protection by using filters to route TCP/IP traffic and NAT (Network Address Translation). An Ethernet router sitting between your DSL modem and your LAN offers the same capabilities of a DSL router.

A *filter* is a rule you establish on the router so that it drops or lets pass certain packets. By establishing a collection of filters within the DSL router, you define its basic security features. Most DSL and Ethernet routers include a minimum set of filters to check data. The three fields that can trigger the filters are the source address, the destination address, and the port number.

Many routers include tools to build your own filters to create more security, but it's a complicated process that demands a high level of expertise. For most small businesses, teleworkers, and consumers, configuring a router to enhance its security isn't practical. You're better off going with a dedicated firewall product in terms of monitoring and filtering TCP/IP packets.

Network Address Translation (NAT) is a common DSL and Ethernet router feature that enables you to create a basic security for your Internet connection. This function converts visible IP addresses and routing provided by your ISP into invisible, non-routed private IP addresses that can't be seen on the Internet. The NAT feature creates a temporary connection between the private IP address and the Internet-routable IP address. The downside of using NAT with private IP addresses is that it can restrict the use of your Internet connection. Because you aren't using registered IP addresses, the computers on your LAN can't be directly linked to the Internet for activities (such as virtual private networking, running servers, or video conferencing).

Most router vendors include a workaround for NAT's inherent limitations for not supporting hosts on a network. These features let you map incoming data traffic to a computer using a private IP address on the LAN behind the router. It uses the Internet-routable IP address assigned to the router as the IP address for the computer using a specific private IP address. This allows you to use VPN, video conferencing, or run an Internet server.

NAT is embodied as a feature also in most other security and Internet-sharing solutions, including proxy servers, security appliances, and Internet-connection-sharing software.

Proxy servers

A proxy's job is to accept requests from a machine on the internal network, screen it for acceptability according to specified rules, and then forward it to a remote host on the Internet. With proxy software and the installation of two NICs, a computer becomes the proxy server for the LAN. One NIC connects to the DSL connection and the other connects to the LAN hub. A variety of proxy server software is available, including Microsoft Proxy Server, WinGate, WinGate Pro, and WinProxy.

In most cases, you can go to the vendor's Web site and download a free evaluation version of the proxy server software to try it out.

Microsoft Proxy Server

Microsoft Proxy Server 2.0 (www.microsoft.com) is a full-featured firewall product that delivers controlled Internet access and monitoring of Web usage. As a BackOffice family member, it's designed for Windows NT and 2000 Server. Microsoft Proxy Server 2.0 lists for $995.

Microsoft Proxy Server also includes a Web cache server to improve the performance of your DSL connection by cutting down on the number of requests the client needs to generate for servers on the Internet. Microsoft Proxy Server includes support for NAT (Network Address Translation).

Proxy Server 2.0 delivers firewall-class security. In addition to being resistant to common attacks, such as IP spoofing, Proxy Server provides packet filtering and access control to block users behind the proxy from accessing certain sites and resources. This feature lets an administrator reject specific packet types at the IP level before they reach higher-level application-layer services.

Enabling packet-filtering causes Proxy Server to drop all packets sent to a destination, except those that match a list of predefined packet filters. You create a filter for the packet types you want the Proxy Server to accept. Microsoft has defined a set of reasonable default packet filters.

Proxy Server can also alert you to suspicious activity at the packet level. For example, if the proxy server rejects more than 20 packets in one second (the default), you're alerted that your network may be under attack.

Proxy Server includes a new feature called dynamic packet filtering. A packet filter does its job based on a TCP service and a port number. For example, a Web server typically uses port 80. A packet filter must always be listening at port 80 for any traffic bound for the Web server. Therefore, the system always has an open port that an intruder could exploit. In dynamic packet filtering, the proxy listens to port 80, but the port is not truly open. When a request is made at port 80 for an HTML document, for example, the proxy opens the port to allow the packet through. As soon as the conversation is over, the proxy closes the port, and the system is locked down again.

WinProxy

For smaller organizations, WinProxy (www.winproxy.com) is a good choice. WinProxy is one of the easiest to install and configure, and it costs a fraction of the price of Microsoft Proxy Server. WinProxy sells for between $99 and $299, depending on the number LAN users, and runs on Windows 95, 98, or NT. WinProxy can operate with either a fixed or a dynamic IP address on the Internet gateway (bridge or router). WinProxy handles most Internet protocols, including POP3, FTP, HTTP, DNS, NTTP, and IMAP4.

WinProxy's HTTP proxy service can support incoming as well as outgoing HTTP requests, allowing you to use WinProxy as an incoming firewall to a Web server located on your LAN. WinProxy lets you restrict Internet access to specific client PCs by IP address, and you can fine-tune Internet access by restricting users to specific protocols. WinProxy's URL filtering feature enables you to block an unlimited number of Internet destinations. This blacklist of taboo Internet sites goes into a text file that the server reads on bootup.

WinGate

WinGate 3.0 (www.wingate.deerfield.com) is an affordable proxy server from Deerfield that runs on Windows 95, 98, and NT. It's flexible, and many shortcomings found in previous versions have been dramatically improved. Improved installation wizards simplify setup and greatly reduce configuration requirements. WinGate 3.0 also includes default security configurations so that users can get their system up and running quickly. WinGate comes in three versions: Home, Standard, and Pro, with the following price breakdowns:

- ✔ WinGate Home costs $40 for 3 users or $70 for a 6 users.

- ✔ WinGate Standard ranges in price from $80 for 3 users to $699 for unlimited users.

- ✔ WinGate pro ranges in price from $299 for 6 users to $949 for unlimited users.

SyGate for Home Office

SyGate for Home Office from Sybergen (www.sybergen.com) ranges in price from $40 for 3 users to $143.94 (how did they come up with this price!) for unlimited users. SyGate for Home Office is designed to be easy to set up and run. It supports Internet sharing with NAT and DHCP, prevents hackers from entering your network with built-in a firewall, and manages Internet access options.

Firewall in a box: Security appliances

An *Internet security appliance* is a firewall in a box designed to provide a turnkey security solution for your LAN. The key benefit of using a security appliance is that security functions are moved off your computers to a box that sits between your DSL CPE and your LAN. These security appliances typically incorporate router functions such as NAT and DHCP as well as VPN support. Most of these devices now work with the latest Internet software, such as video conferencing, online multiplayer games, and streaming audio and video.

A growing number of vendors are entering the security appliance market. Costing several hundred dollars, these security appliances are targeted at small businesses.

SonicWALL (www.sonicwall.com) is the leading firewall appliance vendor and sets the standard for security appliances. Other security appliances that you should check out follow:

- ✔ **WatchGuard SOHO** ($399) from WatchGuard Technologies (www.watchguard.com) is a new security appliance targeted at the same market as the SonicWALL. At the time of this writing, the WatchGuard SOHO wasn't yet released.

- ✔ **WebRamp 700** ($479) is a security appliance from Ramp Networks (www.ramp.com) that is based on the SonicWALL technology.

- ✔ The **OfficeConnect** family of Internet Firewalls from 3Com (www.3com.com) are security appliances based on the SonicWALL technology. The 3Com OfficeConnect Internet Firewall 25 ($695) incorporates the same features of the SonicWALL SOHO but for 25 computers.

More powerful security appliances are available from SonicWALL (SonicWALL Pro) and Netopia (S9500 Security Appliance), but at around $2995 each, these are beyond the reach of smaller businesses. One of the key enhancements of these products is that their beefed-up processing power to support the demands of comprehensive firewall filtering of data traffic, VPN, and other enhanced security features.

Check out the SonicWALL SOHO

SonicWALL makes a line of firewall products that it calls Internet security appliances. The SonicWALL SOHO/10 ($495), shown in Figure 9-1, supports up to 10 PCs, and the SonicWALL SOHO/50 ($995) supports up to 50 PCs. Sporting a compact design — about the size of a VHS tape — with a built-in, four-port hub, the SonicWALL SOHO plugs in between your LAN and the DSL modem.

Figure 9-1:
The
SonicWALL
SOHO.

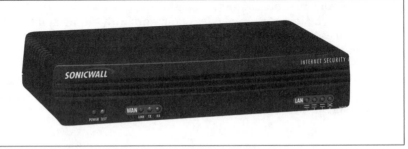

Photo courtesy SonicWALL

SonicWALL SOHO delivers the following impressive features for the small business:

- ✔ **Enterprise-Level security.** The SonicWALL SOHO uses Stateful Packet Inspection, a sophisticated firewall technology found in enterprise-level firewalls. The default settings deliver instant security, and you can easily customize the firewall by adding new services or by creating your own rules using Network Access Rules.

- ✔ **Denial of Service protection.** Right out of the box, the SonicWALL SOHO is configured to automatically detect and thwart denial of service attacks such as Ping of Death, SYN flood, LAND Attack, and IP spoofing.

- ✔ **Java, ActiveX, cookie, and proxy blocking.** Java and ActiveX offer enhancements to Web pages, but hackers can use them to steal or damage data. SonicWALL SOHO can examine HTTP traffic and block the Java and ActiveX portions of a Web page download as well as block cookies. You can also customize the SonicWALL to specify from what trusted sites users can download Java and ActiveX components. In addition, SonicWALL can disable access to proxy servers located on the WAN. (When a proxy server is located on the WAN, LAN users pointing at the proxy server might be able to circumvent content filtering.)

- ✔ **Router-like IP address management.** SonicWALL SOHO includes NAT (Network Address Translation) and DHCP (Dynamic Host Configuration Protocol). It also supports one-to-one NAT to map external IP addresses to private IP addresses hidden by NAT. This allows computers using private addresses to be accessed from the Internet.

- ✔ **Ease of administration.** SonicWALL SOHO has a Web-managed interface, which makes configuration easy using a Java-enabled browser (such as Netscape Navigator). The SonicWALL SOHO installation process uses a handy wizard to guide you through the initial configuration.

- ✔ **Upgradable firmware.** SonicWALL SOHO uses flash firmware that allows you to update its features by downloading any new firmware upgrades from the SonicWALL Web site. You can install a firmware update with a single button click.

- ✔ **User remote access from the Internet**. Using a Web browser, users on the Internet can access an intranet on a private LAN. SonicWALL SOHO uses MD5 encryption to ensure the privacy of all usernames and passwords used to log on for management or remote access.

- ✔ **Security event log.** SonicWALL SOHO maintains a log of events that can monitor security concerns. You can view the log using SonicWALL SOHO's Web-managed interface, or you can be notified by email of an attack on a server.

✔ **Optional IPSec VPN upgrade.** SonicWALL SOHO can be upgraded to support IPSec-based VPN. This VPN solution supports the use of VPN between remote LANs with a SonicWALL SOHO at each office.

✔ **Optional Internet content filtering**. SonicWALL SOHO allows you to create and enforce Internet access policies. This enables you to block incoming content as well as make certain users exempt from filtering. When a user tries to access a forbidden site, you can deny access, create a log, or both. Because Internet sites with offensive material change constantly, you must purchase a subscription service from SonicWALL to maintain content filtering.

Your very own personal firewall

Although personal firewalls don't have the full-strength security of proxy servers or security appliances, they offer an affordable first line of defense for consumers and SOHOs. Personal firewall software installs on each computer connected to your DSL Internet connection. The personal firewall monitors the data traffic coming into your PC from the Internet. If it detects a potential threat, it responds to prevent any damage. Personal firewalls provide protection from a variety of hacker programs and often include other security tools.

The Web sites of most personal firewall vendors include a free evaluation version that you can download, to take the software out for a spin.

ZoneAlarm is freeware for individuals using it for personal use. You can download the program from the ZoneLab site at www.zonelabs.com.

✔ **ZoneAlarm** (free for individuals and non-profit groups; $20 for commercial and government organizations, but you can download it for a 60-day trial period) from ZoneLabs (www.zonelabs.com) works on Windows 95, 98, NT, and 2000. It includes five interlocking security services to deliver easy-to-use and comprehensive protection. ZoneAlarm incorporates a firewall, application control, Internet Lock, dynamically assigned security levels, and zones. ZoneAlarm comes configured, ready to defend against Spyware such as Back Orifice and BackDoor-G. You can also grant and deny access privileges and automatically lock your PC after a given period of inactivity. Figure 9-2 shows the ZoneAlarm window, which gives you access to its five interlocking security services.

✔ **ConSeal Private Desktop** ($50) from Signal9 Solutions (www.signal9.com) is a comprehensive firewall program that works on Windows 95 and 98. McAfee.com acquired Signal9 Solutions in January 2000.

✔ **Sybergen Secure Desktop** ($30) from Sybergen Networks (www.sybergen.com) works on Windows 95, 98, and NT. It includes intruder detection, access security, and access monitoring. You can choose from five security levels to establish the level of protection you want.

✔ **Norton Internet Security 2000** ($60) from Symantec
(www.symantec.com) combines a personal firewall with Norton
AntiVirus software, privacy protection, and Web content filtering.
Norton Internet Security 2000 also includes cookie control to allow you
to specify which Web sites can store cookies on the computer. A *cookie*
is a mechanism that allows a Web site to place information on your computer. The downside of cookies is that they can be placed on your computer without your knowledge to monitor your activities or gather
information.

✔ **BlackIce Defender** ($40) from Network Ice (www.networkice.com)
works on Windows 95, 98, and NT Workstation. BlackIce uses a simple,
user-friendly interface that lets you control the degree of security you
want to implement. Your choice of Confidence levels are Trusting,
Cautious, Nervous, and Paranoid; the default setting is Cautious.
BlackIce shows you the type of attack, identifies the intruder, and provides a real-time graph that tracks network and attack activity based on
time and severity.

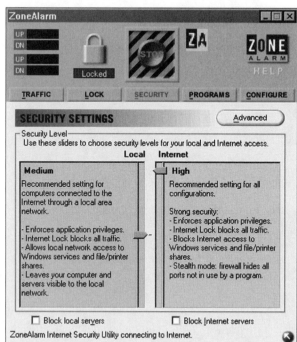

Figure 9-2:
The
ZoneAlarm
window.

Chapter 10

Sharing the DSL Experience

. . . .

In This Chapter

▶ Sharing begins in the home or office

▶ Discovering your DSL Internet-connection-sharing options

▶ Using an Ethernet router route

▶ Checking out EtherFast

▶ Presto! Turning your computer into a router

. . . .

DSL was made for sharing. With ten to fifty times the speed of a dial-up modem, a DSL connection has the power to support multiple computers. Yet today's typical consumer and SOHO DSL offerings include a DSL modem with a single IP address. If you have more than one computer, the first question to ask is "How can I share my DSL service among two or more computers?" This chapter provides the answers to sharing your DSL connection.

Assembling a LAN — Don't Panic

DSL modems connect to the DSL line through the telephone line and to your computer or LAN at the other. The underlying technology for sharing a DSL Internet connection among multiple computers is Ethernet. Ethernet forms the basis of the most popular form of computer networking for PCs and Macs.

DSL CPE can connect to your computer using Ethernet, USB, or PCI. Using an Ethernet DSL modem gives you a choice of three ways to share your DSL Internet connection. You can share USB or PCI DSL modems by using the computer connected to the USB or PCI modem as the gateway for all the computers on a network.

The first order of business in sharing an Internet connection is setting up a LAN. Don't panic! Today, assembling the basic LAN plumbing of network interface cards (NICs), cabling, and hub is easy and inexpensive. And it's getting better in terms of getting even easier to install, falling prices, and new technologies, such as USB adapters, Phoneline, and wireless LAN. See Chapter 11 for specifics on setting up your own LAN.

Share and Share Alike

Four technologies are available for sharing an Internet connection that uses a DSL modem and a single IP address. These options follow:

- ✔ A **proxy server** that combines the NAT and DHCP router functions for connection sharing with session-level security.

- ✔ A **security (firewall) appliance** that includes built in NAT and DHCP. The device sits between your DSL modem and LAN.

- ✔ An **Ethernet router** that sits between your DSL modem and LAN and connects to each through Ethernet. This router, sometimes referred to as an Ethernet-to-Ethernet router, includes NAT, DHCP, and other router features used by a DSL router. What an Ethernet router can't do is connect directly to the DSL line.

- ✔ **Internet-sharing software** delivers the connection sharing of a proxy server but without the security overhead. This solution uses a computer as the router for your LAN.

This chapter focuses on Ethernet routers and Internet-connection-sharing software.

Proxy servers and security appliances

Proxy servers and security appliances are heavily intertwined with security solutions. Especially in proxy servers, security features can add a layer of complexity to the task of sharing the connection. But proxy servers can complicate Internet sharing by restricting incoming traffic to the exclusion of some of the newer, popular Internet applications such as instant messaging, video conferencing, games, and streaming audio and video. For more on proxy servers and security appliances, see Chapter 9.

Ethernet routers and sharing software

Ethernet routers incorporate most of the features of a DSL router but without the DSL line connection. Instead, the Ethernet router sits between the DSL modem (bridge) and the LAN made up of the computers you want to share.

Internet-sharing software turns a PC into a router. You install two NICs (network interface cards) in the PC, and the Internet-connection-sharing software creates the router to share the connection.

Going the Ethernet Router Route

An Ethernet router is an external box that sits between the DSL modem and your LAN. It has two Ethernet ports. Don't confuse an Ethernet router with a DSL router. An Ethernet router connects between a DSL modem and computers — not to the DSL line. You connect the 10BaseT cable from your DSL modem to one port and the 10BaseT cable from your LAN hub to the other port. Most Ethernet routers include a built-in hub to jumpstart your network. All you need are the NICs and cabling.

Ethernet routers enable sharing through NAT and DHCP, which also adds some protection to your DSL modem connection. One-to-many NAT doesn't support dynamic port mapping for streaming protocols used for multiplayer games, video conferencing, and other multimedia applications, but most Ethernet routers support the use of these applications anyway.

DSL CPE vendors offer full-featured Ethernet routers designed for the small business market, with prices of $500 and up. Unfortunately, most of these routers are expensive, and nontechnical users will find their setup complicated.

Following are two DSL CPE vendors offering Ethernet routers:

- **Netopia R9100** Ethernet router from Netopia (`www.netopia.com`)
- **Cayman 2E500** and **Cayman 2E500** (with built-in hub) Ethernet routers from Cayman Systems (`www.cayman.com`)

Heavy competition among a growing number of vendors means that you can expect prices to continue to fall. The under-$200 price barrier for an Ethernet router has already been broken for the SOHO and consumer market.

The following Ethernet routers are targeted at the SOHO and consumer markets:

- **UGate 3000** ($399) from UMAX (`www.umax.com`).
- **EtherFast Router** ($199) from Linksys (`www.linksys.com`).
- **ISB2LAN** ($349) from Nexland (`www.nexland.com`).
- **Internet Access Gateway Router** ($350) from NetGear (`www.netgear.com`).
- **X-Router** ($299) from MacSense (`www.macsensetech.com`).

Checking Out the EtherFast Router

Linksys, a leading PC networking hardware manufacturer, has produced the first truly consumer router. See Figure 10-1. The EtherFast Router represents the state-of-the-art in Internet-sharing devices in terms of performance, ease of use, and price. Sporting a cool black-and-blue, stackable box at a great price ($155 street), the EtherFast router includes impressive features that rival routers costing hundreds of dollars more. Designed for the nontechnical user, the EtherFast Router is an excellent consumer choice for sharing a single IP Internet connection.

Figure 10-1:
The Linksys
EtherFast
router.

Photo courtesy Linksys

Built-in switch

The EtherFast router sits between the DSL modem and a LAN. Don't have a LAN? Not a problem. The EtherFast router includes a built-in 4-port 10/100 switch to build a network around it. A switch improves network performance over a hub because it segments a network to squeeze out extra throughput. Instead of passing data around the entire network the way a hub does, the switch efficiently manages data traffic between specific ports. The router also includes an Uplink port to connect another hub to expand the network.

Easy configuration

Configuring the EtherFast ROUTER is amazingly easy using any Web browser. Only four IP address fields need to be configured to begin using the router. These fields, shown in Figure 10-2, include the IP address assigned by the ISP, the subnet mask, the default gateway, and the DNS server address. The DHCP (Dynamic Host Configuration Protocol) server is already configured for 50 computer and ready to go. After the EtherFast router is ready, you simply configure the clients for dynamic IP addressing.

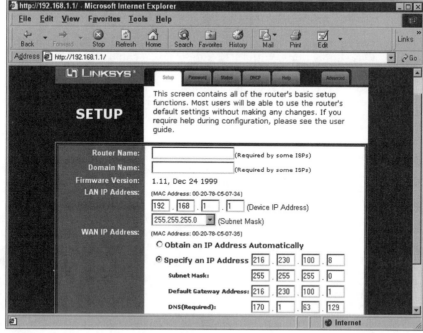

Figure 10-2:
The easy
Web inter-
face for
configuring
the
EtherFast
router.

Beyond NAT

To support Internet sharing, the EtherFast router uses NAT (Network Address Translation) but includes enhancements to overcome the inherent limitations of NAT. Using a feature called Application Sensing Tunnel, the router senses an application that uses dynamic port mapping and automatically opens a multiport tunnel to allow it to pass through NAT. This allows any computer connected to the EtherFast router to use applications that NAT normally won't let through, such as video conferencing, games, and other multimedia applications.

Exposed in the DMZ

The EtherFast router includes a handy DMZ/Expose host feature that allows one PC on the LAN to be a host on the Internet to provide unrestricted two-way communications. This enables you to use any application on a desig-nated PC that requires an Internet-routable IP address, such as video conferencing, VPN, or a personal Web server. The EtherFast router translates its Internet-routable IP address (such as 216.216.6.8) to the private IP address assigned to the PC on the local network (such as 192.168.1.101).

Other noteworthy features

The EtherFast router supports Point-to-Point Tunneling Protocol (PPTP) for VPN connections. With PPTP support, you can use the Microsoft VPN client included in Windows 95, 98, NT, or 2000 to connect to an office network that uses a DSL router such as the Netopia R7100, which includes a built-in PPTP server. Wait, there's more. You can upgrade the router easily by downloading firmware updates from the Linksys Web site, thus preserving your investment.

Making Your Computer a Router

Internet-sharing software turns a PC into a router to support sharing of the DSL connection. The computer acts as a router in the background; you can still use the computer as you normally would. Most Internet-connection-sharing software requires that you install two NICs in the computer running as the router. The All Aboard product, however, requires only one NIC.

If you're using a USB or PCI DSL modem, the PC treats the modem as a network device. This means the software driver for the modem appears as a NIC in the computer. Because the USB or PCI modem connects to the DSL line, the NIC you install in the PC connects the LAN hub. The sharing software creates the router between the two network devices.

Following are your Internet-connection-sharing software choices:

- **WinRoute Home** ($49), **WinRoute Lite** ($149), and **WinRoute Professional** ($199 to $719) from Tiny Software (www.tinysoftware.com) deliver the full features of a router plus security enhancements, an Internet email server, a cache, and a URL filter. They run on Windows 95, 98, NT, and 2000.

- **Microsoft Internet Connection Sharing** (ICS) from Microsoft (www.microsoft.com) is included in Windows 98 Second Edition. Using two NICs, you can create a basic Internet-sharing router with minimal features. Windows 2000 also includes Internet-connection sharing.

- **All Aboard Standard Edition** ($40 to $200) and **All Aboard Business Edition** ($275 to $950) from InternetShare (www.internetshare.com) are the only Internet-connection-sharing software that requires a single NIC. The NIC connects to the hub where the DSL modem and other computers connect. All Aboard works on Windows 95, 98, NT, and 2000.

- **CyberMiser Connect (CMC)** ($299) from CyberMiser (www.cibermiser.com) is connection-sharing software that's easy to set up. CMC includes a simple firewall and an Internet email server.

If you're interested in a particular brand of Internet-connection-sharing software, check out the vendor's Web site for a free evaluation version. That way, you can try it before you buy.

Part III

DSL Internet Service Meets Your Computer or LAN

The 5th Wave   By Rich Tennant

"Ooops — here's the problem. Something's causing shorts in the server."

In this part . . .

This part takes you to where the DSL Internet connection meets your individual PC or Mac or a local area network (LAN). For those who want to connect multiple PCs and Macs to a DSL connection, you find out how to set up the basic LAN plumbing for the three leading network types: Ethernet, Phoneline, and wireless. After you have a LAN in place, you have the foundation for sharing your DSL Internet service.

As part of making your PC or LAN connection to the Internet through DSL, you walk through configuring Microsoft TCP/IP for Windows 95/98, Windows NT, Windows 2000, and Mac computers.

For those using an Ethernet DSL modem or router to connect to the Internet, you find out how to configure the TCP/IP settings for your NIC (network interface card). I also show you how to work with Microsoft Windows Dial-Up Networking to support certain types of PCI and USB DSL modem service. Finally, you'll find out how configure TCP/IP for the latest Macs.

Chapter 11

Building a Network to Share DSL

. . . .

In This Chapter

▶ Sharing a DSL connection using a network

▶ Defining your LAN plan

▶ Exploring your Ethernet, Phoneline, and wireless LAN options

▶ Finding out about NICs

▶ Connecting a NIC to your computer

▶ Working with network cabling

▶ Understanding LAN hubs and switches

▶ Networking Macs

. . . .

*H*igh-speed, always-on DSL Internet service is ideal for sharing across two or more computers. At the heart of sharing a DSL Internet connection is a local area network (LAN). The LAN connects the computers in your home or business to share network resources, including the DSL Internet connection.

Today, setting up basic LAN plumbing is easy and inexpensive. And you can choose from a variety of networking technologies to fit your needs, including Ethernet, Phoneline, and wireless networking solutions. This chapter explains how to assemble a LAN as the foundation for leveraging your DSL connection.

Sharing DSL with a Network

If you have more than one computer in your home or business and plan to connect them to your DSL Internet connection, you need a network. A local area network enables you to share information locally, share resources such as a printer, and share your DSL Internet connection.

Setting up a network doesn't mean you can immediately share your DSL connection. Sharing DSL Internet service is intertwined with your IP address configuration and the DSL CPE. The following describes how Ethernet, USB, and PCI DSL CPE interface to a LAN:

- ✔ If you're using **a consumer DSL offering bundled with an Ethernet DSL modem (bridge) and a single IP address,** you need to add an Ethernet router or another connection-sharing solution to connect your LAN to the DSL connection.

- ✔ If you're using **a DSL router,** you can share the DSL connection using a single IP address or you can use multiple routable IP addresses assigned to the computers on your LAN. You have everything you need to share your connection.

- ✔ If you're using **a PCI or USB DSL modem with a single IP address,** you also need to add an Internet-connection-sharing solution. To share a PCI or USB DSL modem connection, you need to use either Internet-sharing or proxy server software and install two NICs in the gateway computer (the computer connected to the PCI or USB modem).

- ✔ If you're using **an Ethernet DSL modem with multiple IP addresses**, you don't need to use any Internet-sharing solution because each computer on your LAN has its own IP address.

Even if you're DSL Internet service uses a dynamic IP address assigned by your ISP, you can still share the DSL connection. Most Internet-connection-sharing solutions support the use of static IP or dynamic IP addressing.

See Chapter 7 for more information on IP addressing issues, Chapter 8 for more information on DSL CPE options, and Chapter 10 for more information on Internet-sharing options.

DSL and LANs: A symbiotic relationship

The correlation between broadband and networking is symbiotic. Jupiter Communications estimates that by the end of 2000, the total number of broadband customers will reach 4 million, with just a slightly smaller number of home networks. By 2003, there will be 15 million broadband users and 16 million households with a network — a 400 percent growth in the number of home networks.

The Yankee Group estimates more than 17 million U.S. households, about 37 percent of all households with PCs, are interested in home networking. And nearly 50 percent of households with multiple PCs reported that they're considering a home network. Households aren't the only ones facing the network need; nearly 50 percent of the less technologically sophisticated small businesses and offices don't have a network.

Your LAN Plan

Building a network from the ground up involves adding network interface cards (NICs) to your computers and connecting them with cabling or using a wireless connection to each other or to a network hub. This forms the basic hardware infrastructure for your network and the foundation for connecting the DSL modem or router to your LAN.

After the LAN hardware in place, you configure the workstations for networking using the network operating system (NOS). For specific information on configuring

- ✔ Microsoft Windows 95 or 98 for TCP/IP networking, see Chapter 12
- ✔ Microsoft Windows NT for TCP/IP networking, see Chapter 13
- ✔ Windows 2000 for TCP/IP networking, see Chapter 14
- ✔ Your Mac for TCP/IP networking, see Chapter 15

Do-it-yourself networking

Networking solutions targeted at the home, SOHO (small office/home office), and small business markets are inexpensive and easy to set up. Networking companies have refined their networking products to make self-installation possible. They also offer customers choices in networking technologies, including Ethernet, Phoneline, and wireless networking solutions. Network vendors have also packaged their products into complete Ethernet or Phoneline kits for assembling small networks, with most of these kits costing less than $100.

You can choose products from a number of PC network vendors, including the following:

- ✔ 3Com (www.3com.com)
- ✔ Apple — Mac (www.apple.com)
- ✔ Asante (www.asante.com)
- ✔ DLink (www.dlink.com)
- ✔ Farallon — Mac (www.farallon.com)
- ✔ Intel (www.intel.com)
- ✔ Linksys (www.linksys.com)
- ✔ MacTech, (www.maxtech.com)
- ✔ NetGear (www.netgear.com)

Power Networking: Home networking through electrical wires

Home networking over power lines has an irresistible upside: Power lines and wall sockets are ubiquitous in most homes. Home networking over power lines is in its developmental stage — you won't see any serious home networking over powerline products until the first half of 2001 (by some estimates). Momenteum for powerline networking, however, does exist: Microsoft and Intel have bought companies that deal with powerline networking.

You can typically find all the networking hardware you need at your local computer retailer or on the Internet at any number of sites that sell computer hardware.

If you don't want to install your own LAN, check out TechPlanet by visiting www.techplanet.com or calling 877-832-4752. TechPlanet is nationwide technology service company that does on-site network and other technology installations and provides support specifically for small businesses.

Your networking choices

You have three choices in the type of networking technologies you can use to connect your computers: Ethernet, Phoneline, and wireless.

The most popular form of networking in use today is Ethernet. Phoneline networking is a newer technology designed to network computers using existing telephone wiring. Wireless networking is even newer and uses radio frequencies to connect your computers.

The following briefly describes these networking options, which are explained in more detail later in this chapter.

- ✔ **Ethernet** forms the basis of most PC and Mac networking and comes in two flavors: standard Ethernet (10 Mbps) and Fast Ethernet (100 Mbps). Ethernet is a popular interface for DSL CPE. An Ethernet DSL modem or router connects to a single computer's NIC or to a hub or a switch for a LAN.

- ✔ **Phoneline networking,** as its name implies, uses your telephone wiring as network cabling. Network data traffic runs over telephone wiring used by telephones, fax machines, analog modems, and ADSL/G.lite modems without interrupting these services. Phoneline networking is not Ethernet-based and requires a bridge to connect to Ethernet DSL CPE.

✔ **Wireless networking** is the newest networking technology and is going through the growing pains of competing specifications. Second-generation wireless networking products support 11 Mbps (about the same speed as standard Ethernet).

Lay of the LAN

In networking jargon, *topology* refers to the layout of the network. Two basic network topologies are used for PC networks: star and bus, or daisy chain.

A *star* topology uses a hub or a switch as the center of the network. A *hub* is a simple device that acts as an interconnection point for the computers connected to the LAN. Each computer is connected to the hub using a cable that runs from the NIC to the hub. A *switch* plays a similar role as a hub except it includes more intelligence to better manage network data traffic. One of the best features of the star configuration is that damage to any given cable is likely to affect only a single computer, not the whole network.

A bus, or daisy chain, topology is basically a straight line of cable that runs from one computer to another in a linear fashion. Although you can route the cable in a twisted and circular arrangement, the two ends of the cable are never connected.

The three networking technologies use the following topologies:

✔ Most Ethernet networks today use the star topology. Some older networks use a bus topology, which uses coaxial cabling called 10Base2. Fast Ethernet works only with the star topology. Figure 11-1 shows an Ethernet star network configuration. The 10BaseT network cables radiate from a central hub or switch to every computer or other network device on a network.

✔ Phoneline networks use a daisy chain topology. You connect one end of the telephone wire to the Phoneline NIC's RJ-11 telephone port in the first computer, and connect the other end of the cable into a telephone jack in your wall. You can connect computers in different rooms to the telephone wall jacks in each room. You can also connect multiple computers to a single telephone wall jack by running telephone wiring from the first computer to the second computer and so on for each computer on your network. Figure 11-2 shows a Phoneline daisy chain topology.

✔ Wireless networks can be either daisy chain or star topologies. Home Radio Frequency Working Group (HomeRF) networks connect PCs without the use of a hub device. Wireless networks based on the IEEE 802.11b High Rate (HR) specification use a star topology with the wireless hub called an *Access Point*. The Access Point also acts as a bridge between Ethernet and wireless networks.

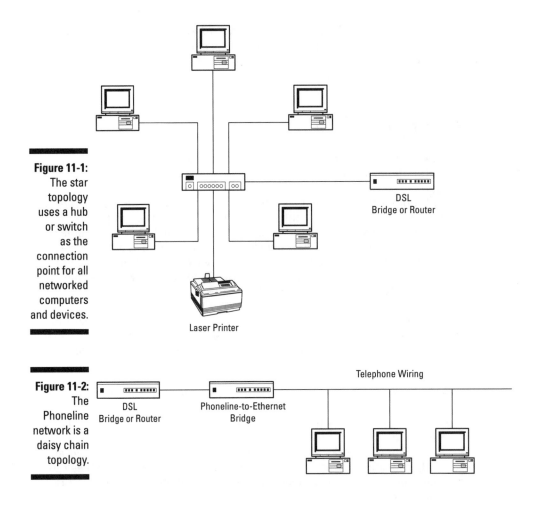

Figure 11-1: The star topology uses a hub or switch as the connection point for all networked computers and devices.

DSL Bridge or Router

Laser Printer

Figure 11-2: The Phoneline network is a daisy chain topology.

Telephone Wiring

DSL Bridge or Router

Phoneline-to-Ethernet Bridge

Tried and True Ethernet

Ethernet forms the basis of most PC and Mac networks. There are two Ethernet specifications: standard Ethernet (10 Mbps), which is based on the IEEE (Institute of Electrical and Electronics Engineers) 802.3 standard, and the newer Fast Ethernet (100 Mbps), which is based on the IEEE 802.3u standard.

Because both forms of Ethernet are based on the IEEE 802.3 standard, they're mutually compatible. This means you can integrate Fast Ethernet hardware into an existing Ethernet network. In fact, most NICs sold today are 10/100 adapters that support both 10 Mbps and 100 Mbps to accommodate mixed

10-Mbps and 100-Mbps networks. To use the greater speed, your network needs Fast-Ethernet-compatible hubs and switches. Newer autosensing 10/100-Mbps hubs and switches can detect both forms of Ethernet and make adjustments on the fly.

If you don't have a network already in place, go for Fast Ethernet NICs and hubs and switches. The cost differential between 10-Mbps Ethernet and 100-Mbps Fast Ethernet equipment is inconsequential. By using Fast Ethernet, you enable users to do more things — now and in the future — on both the local network and the Internet.

Most Ethernet NICs sold today are PCI adapter cards that support both Ethernet and Fast Ethernet. USB (Universal Serial Bus) Ethernet adapters are also available that connect to a PC through a USB port.

Ethernet and Fast Ethernet networks use a star topology that includes a hub or switch at the center, an Ethernet adapter in every computer or other network device, and 10BaseT cabling to connect them together. 10BaseT cabling looks like a thicker version of telephone wire; it has eight wires instead of four and uses a larger connector than the four-wire version.

The biggest drawback to using Ethernet is the cabling. Most homes and small offices are not wired for Ethernet networking, which means the customer would need to install 10BaseT cabling.

If it's connected, it has a MAC address

Each node on an Ethernet network is assigned a unique address called a MAC (Media Access Control) address. This is the protocol that controls access to the physical transmission medium on an Ethernet LAN. The MAC layer is implemented in a NIC or any other Ethernet device connected to the network, such as a DSL bridge or router. Every Ethernet device has a unique MAC address, such as 00-80-AE-00-00-01.

MAC addresses play an important role in Ethernet networking and in the way DSL bridges work. Bridges work at the MAC layer, which means they work with MAC addresses to route data. Routers, on the other hand, work at the IP (Internet Protocol) layer to route data.

The number of computers supported by a given DSL bridge is controlled by the number of MAC addresses that the bridge can recognize. A single-user DSL bridge recognizes data traffic from only the specific MAC address it was configured to accept. Multiuser bridges have a database of MAC addresses that they use to identify each Ethernet device on the LAN.

A typical multiuser DSL bridge learns all the MAC addresses of all the NICs on the local network. That way, the bridge can determine whether the data it receives is intended for the local network or for the Internet.

What's My Line? Phoneline Networking

The Home Phoneline Networking Alliance (HomePNA) created the specification for Phoneline networking, which uses — you guessed it — telephone wiring as network cabling. Network data traffic runs over telephone wiring used by telephones, fax machines, analog modems, and ADSL/G.lite modems without interrupting these services.

Most network vendors now offer HomePNA-based networking solutions that support 10 Mbps. The first generation of Phoneline networking supported only 1 Mbps. You install Phoneline NICs in PCs in the same way that you install an Ethernet adapter, but they connect to a telephone line instead of to 10BaseT cabling.

Phoneline networking uses the daisy chain topology, which links computers to each other without the use of a hub. One computer on the network acts as a host for sharing resources and devices on the network. Your Phoneline NIC appears as a network adapter in your computer's operating system (OS). After you install your Phoneline NICs and connect your computers together, you use the computer's OS to share your resources.

Because Phoneline networking isn't an Ethernet-based form of networking, sharing a DSL connection requires the use of Internet-sharing software, proxy software, or a Phoneline-to-Ethernet bridge. If you use Internet-sharing or proxy server software, you must install two NICs on the PC that acts as the gateway to your DSL connection.

A Phoneline-to-Ethernet bridge is a small box that enables you to connect your Phoneline network in one port and an Ethernet network or Ethernet DSL CPE in a second port. Figure 11-3 shows the Linksys HomeLink Broadband Network Bridge is a Phoneline-to-Ethernet bridge.

Figure 11-3:
The Linksys
HomeLink
Broadband
Network
bridge.

Photo courtesy Linksys

Going Cordless: Wireless Networking

As *Wired* magazine put it back in 1997, "To be truly wired, you must be wireless." Wireless networking promises to bring a new level of flexibility and mobility to computer networking for homes and offices.

Wireless networking is the newest member to the networking realm, and is in a confusing state because two incompatible technologies are fighting for supremacy: HomeRF and WECA.

The Home Radio Frequency Working Group (HomeRF), with its SWAP (Shared Wireless Access Protocol) specification, uses the 2.4-GHz band. HomeRF delivers 1- to 5-Mbps data transmission speeds as well as support for four high-quality voice transmission channels. Cayman Systems, Compaq, Intel, and Proxim offer HomeRF products.

The Wireless Ethernet Compatibility Alliance (WECA) backs the IEEE (Institute of Electrical and Electronics Engineers) 802.11b High Rate (HR) specification. This specification, which was ratified by the IEEE in 2000, operates in the 2.4-GHz spectrum and supports a range of speeds from 1 Mbps to 11 Mbps. Compliant IEEE 802.11b wireless equipment displays the Wi-Fi certified logo.

HomeRF networks use a topology that links computers without the use of hub. However, to share a DSL connection through a HomeRF network, you need to use Internet-sharing or proxy server software.

802.11b networks use an Access Point, which acts a hub for the wireless NICs and a bridge for connecting to a wired Ethernet network.

To have access to network services — such as Internet access through the DSL CPE, printing, and network-based applications — the wireless LAN must be connected to a wired Ethernet backbone through the Access Point. The Access Point includes a bridge and a 10BaseT port so that the wireless network can connect to the Ethernet network through the network hub. Figure 11-4 shows the topology for an 802.11b wireless LAN connected to a wired Ethernet network. This topology allows wireless computers to print to a networked printer and the DSL connection through the Ethernet DSL modem.

All wireless networks can suffer from interference from microwaves and cordless 2.4-GHz telephones. And in most cases, actual throughput is less than wired Ethernet. Wireless networks have a maximum range of 300 feet, but the actual range is dependent on the wireless network's environment. Obstacles such as cement walls, steel beams, and elevator shafts can interfere with the radio waves.

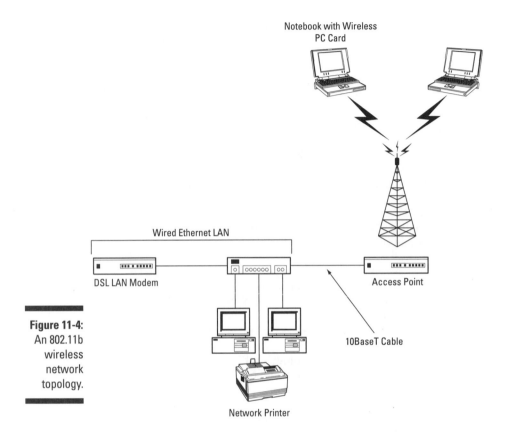

Figure 11-4:
An 802.11b wireless network topology.

Wireless networking is considerably more expensive than Ethernet or Phoneline networking. A typical wireless PCI NIC costs between $99 and $349. The cost for an Access Point is between $299 and $1995.

Wireless networking products are available from a variety of sources, including the following:

- 3Com (www.3com.com)
- Cisco (www.cisco.com)
- Lucent (www.lucent.com)
- Nokia (www.nokia.com)
- Proxim (www.proxim.com)
- RadioLAN (www.radiolan.com)
- WebGear (www.webgear.com)

What's a Bluetooth?

Bluetooth is a specification for short-range, low-power wireless linking of pagers, cell phones, personal access devices, and notebooks. It allows these devices to connect to fast access points, for example, a DSL Internet connection inside your home or in a public place such as an airport. Because of the short range and low power of Bluetooth, it's not designed for the demands of a LAN.

NIC Your Computer

Any device that wants to communicate on a network must have a network interface. A network interface card, or NIC (pronounced "nick"), is also referred to as a network controller. A network interface might come built into a computer, or you might need to install it yourself as an adapter card. For connecting computers to an Ethernet, Phoneline, or wireless LAN, you need a NIC.

NICs come in a variety of flavors in terms of the way that they interface with your computer. The three leading interfaces used by NICs are

- ✔ PCI (Peripheral Component Interconnect)
- ✔ PC Card
- ✔ USB (Universal Serial Bus)

ISA (Industry Standards Association) is a slower 16-bit bus used by older computers. Although many of today's computers include one or two ISA slots to support older adapter cards, it doesn't make sense to use ISA NICs unless you're trying to connect a PC without any available PCI slots to a LAN.

Ethernet and Phoneline NICs come in all interface flavors (PCI, PC Card, USB); wireless NICs typically come in PCI and PC Card flavors.

Most 10/100 Ethernet and Phoneline NICs retail between $30 and $75. Wireless NICs are typically more expensive than Ethernet or Phoneline NICs, depending on the type of wireless network and vendor. You can expect to pay between $99 and $299 for a wireless NIC; as the technology matures, however, prices will undoubtedly come down.

PCI NICs

Most NICs for desktop computers use the PCI (Peripheral Component Interconnect) bus to connect to your PC. PCI supports 32-bit and 64-bit data paths, which dramatically speeds up a NIC's communications with your PC.

Intel developed PCI to provide a high-speed data path between the CPU and up to ten peripherals while coexisting with older buses, such as ISA.

Using PCI network adapters has other benefits besides LAN speed. The PCI data bus also provides a bus mastering technique that allows more processor independence. This technique reduces CPU overhead by taking control of the system bus, which enables the PC to support more bandwidth-intensive applications.

The PCI bus architecture supports bus mastering and concurrency. In *bus mastering,* an intelligent peripheral, such as a NIC, takes control of the bus and accelerates high-throughput, high-priority tasks without the need for processor intervention. *Concurrency* allows the processor to operate simultaneously with the bus mastering devices and work on other tasks. Most of today's PCs include PCI slots, but only a PCI slot that supports bus mastering can be used with a NIC that supports bus mastering. For example, 3Com's Fast Ethernet cards use a bus mastering architecture. In most PCs, slots 1 and 2 are bus mastering slots. If you're using a video PCI card, it's probably using one of the bus mastering slots.

Figure 11-5 shows the 3Com Fast Ethernet PCI NIC. Notice the RJ-45 port. You connect the NIC using a 10BaseT cable. The cable connects to a hub or directly to the Ethernet DSL modem or router. Figure 11-6 shows a Linksys Phoneline NIC, which has two RJ-11 ports.

Figure 11-5:
A 3Com
Fast
Ethernet PCI
NIC.

Photo courtesy 3Com Corporation

Figure 11-6:
A Linksys
Phoneline
PCI NIC.

Photo courtesy Linksys

PC Card NICs

You can add a notebook computer to a LAN with ease using a PC Card (formerly called PCMCIA), which is a credit-card-size adapter card. You can also get dual-purpose Ethernet and analog modem PC Cards that support both networking and mobile communications. Figure 11-7 shows a Linksys Fast Ethernet PC Card. A Phoneline PC Card looks like the Ethernet PC Card but includes two RJ-11 ports instead of the single RJ-45 port. A wireless PC Card, shown in Figure 11-8, lacks any ports and instead includes a small built-in antenna.

Figure 11-7:
The Linksys
Fast
Ethernet
PC Card.

Photo courtesy Linksys

Photo courtesy Cisco

USB NICs

USB ports have been standard on most PCs for the last few years; Microsoft Windows 98, Windows 2000, and the latest Macs support USB. Windows 95 and Windows NT do not include native support for USB.

A USB NIC is an external connector that includes a USB cable on one end and an Ethernet RJ-45 port or a Phoneline RJ-11 port on the other. At the time of this book's writing, there were no USB NICs for wireless networks. Figure 11-9 shows a Linksys USB NIC.

The beauty of using a USB NIC is that it's easy to install. You simply snap the connector into your computer or USB hub; no case cracking is required. The downside of USB NICs is that they support only 10-Mbps Ethernet due to the inherent speed limitation of USB specification 1.1, which supports data speeds up to 12 Mbps. The new USB specification 2.0, however, will remove the 12-Mbps barrier and support a data speed of up to 480 Mbps.

Figure 11-9:
A Linksys
USB NIC.

Photo courtesy Linksys

Installing NICs

The way you install the NIC in your PC depends on whether you're using a PCI, a PC Card, or a USB NIC. These days, installation is easy — regardless of which type of NIC you're using.

Here's the typical installation process for each type of NIC:

✔ For a **PCI NIC,** which can be for Ethernet, Phoneline, or wireless networks, you must open the PC's case and install the NIC in an available PCI slot.

✔ For a **PC Card,** you insert the credit-card-size NIC into an available PC Card slot on your notebook.

✔ For a **USB NIC,** you connect a cable from the NIC to the USB port on the PC or to a USB hub.

After you install the NIC, the computer's operating system handles the installation of the supporting software driver. The software driver allows the operating system to communicate with the NIC.

The process for setting up a network adapter follows these three main steps, regardless of whether the adapter is a PCI NIC, a PC Card, or a USB NIC.

1. **Install the NIC by connecting it to your computer.**

2. **Install the software driver so that the operating system will recognize the NIC.**

   Driver software is provided on the disk that comes with the NIC.

3. **Bind the networking protocol, such as TCP/IP, to the NIC.**

   Chapter 12 explains how to bind and configure Microsoft Windows 95 and 98 for TCP/IP networking. Chapter 13 does the same for Microsoft Windows NT, and Chapter 14 covers Windows 2000. For Macs, see Chapter 15.

Installing a PCI NIC

Most NIC cards come with instructions and software for installing them in machines with different operating system. In this section, I outline a typical PCI NIC adapter installation. Depending on the NICs you're installing, the installation process might be different.

 Watch out for static electricity. Carry the NIC in the antistatic material it came in, and always ground yourself before you put your hands in a machine or handle computer hardware. To ground yourself to dissipate static buildup, simply touch the chassis of the computer before doing anything else.

Follow these steps to install a PCI NIC in your PC:

1. **Turn off your PC and any peripheral equipment attached to it.**

2. **Remove your PC's case.**

   Each computer manufacturer has a different case design, but in general you remove the screws from the back of the computer case and then slide the case away from the chassis.

3. **Remove the back-panel cover for the free slot you want to use for the NIC.**

   A PCI NIC requires a master PCI slot. In new PCs, PCI slots are all master slots. Older PCs might have both master and slave slots. If you're installing the NIC into an older PC, you need to determine the master PCI slot before installing the NIC — check your computer's documentation or contact the vendor.

4. **Carefully insert the PCI network card into your PC's slot. Make sure all of the card's pins are touching the slot's contacts. Then secure the card's fastening tab to your PC's chassis with a mounting screw.**

5. **Replace your PC cover.**

6. **Connect one cable link for each PC.**

   Plug one RJ-45 connector into the network adapter card at the back of your PC, and plug the other connector into an available port on your hub.

7. **Turn on your PC.**

Microsoft Windows starts, and you begin the process of installing the software driver for the NIC so that the operating system recognizes the installed NIC.

Adding multiple NICs on a PC

You typically use two NICs to share an Internet connection using a proxy server or Internet-connection-sharing software. Microsoft Windows 95, 98, NT, and 2000 support the use of multiple NICs. You can connect additional NICs in the same way you did the first NIC.

If you use a USB or PCI DSL modem and want to share the DSL connection, you need to install an additional NIC. As explained in Chapter 8, the USB or PCI DSL modem is installed as a NIC, so it counts as one of the two NICs you need to share the Internet connection.

Installing the NIC driver

Installing NICs from Windows 95, 98, and 2000 is easy thanks to Plug-and-Play (PnP), which automatically detects any new hardware. After you install the NIC in the computer and turn on the PC, Windows 95, 98, or 2000 detects the adapter card and prompts you for the software driver. Just follow the instructions on the screen to install the software driver.

Installing a NIC driver in Windows NT 4.0 is handled differently than in Windows 95, 98, and 2000 because NT lacks Plug-and-Play support. When you start Windows NT 4.0 after installing the NIC, there is no hardware detection and no Add New Hardware Wizard. You install a NIC driver in Windows NT using the Network properties in the Control Panel.

Where the NIC driver lives in Windows

In Windows 95, 98, and NT, the Network properties dialog box displays your installed NICs. From the Network properties dialog box, you can configure your NIC for TCP/IP to connect to the Internet. The layout of the Network properties dialog box is different in Windows 95 an 98 and Windows NT, but you can access it the same way in all three operating systems: Simply right-click the Network Neighborhood icon on the desktop and then choose Properties from the menu.

In Windows 2000, the Network properties dialog box is called the Local Area Connection Properties dialog box. To get to it, simply right-click the My Network Places icon on the Windows 2000 desktop. Then, in the Network and Dial-up Connections window, right-click the Local Area Connection icon and choose Properties from the pop-up menu.

The Media Is the Message

Cabling is the medium that connects your LAN. In the star topology, cabling runs from every computer (or other network device) to a hub or a switch. Network cabling used by most Ethernet networks is readily available, inexpensive, and easy to install. Phoneline networking cable is ordinary 4-wire telephone wiring. This section explains the fundamentals of Ethernet network and Phoneline cabling.

Know your connectors

At the end of your wiring are connectors. Phoneline networking uses the RJ-11 modular connector, which snaps into an RJ-11 port. This is the same connector used in most homes and small businesses for telephones, faxes, and modems. Larger businesses have PBX-style voice telephone networks that use RJ-45 connectors, which are the same connectors used in 10BaseT cabling for Ethernet networks. Figure 11-10 shows the RJ-11 and RJ-45 connectors.

Figure 11-10:
The 4-wire
RJ-11 and
8-wire RJ-45
modular
connectors.

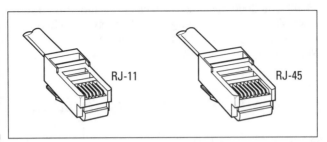

10BaseT cabling

10BaseT cabling is the media used in star topology Ethernet and Fast Ethernet networks. 10BaseT cabling is Category 5 cable, which gets its name from a cable rating system used by the Electronic Industry Association/Telecommunications Industry Association (EIA/TIA).

10BaseT cabling is also referred to as *twisted-pair (TP) cabling*. 10BaseT cabling consists of four pairs of copper wires that are twisted around each other to cancel out interference, hence the name twisted-pair. Ethernet uses four wires (two pairs) of the eight-wire capacity of twisted-pair cabling.

Twisted pair cabling comes in two flavors: shielded twisted pair (STP) and unshielded twisted pair (UTP). UTP is by far the most popular. UTP consists of pairs of copper wires twisted around each other and covered by plastic insulation. STP has a foil or wire braid wrapped around the individual wires to provide better protection against electromagnetic interference. STP uses different connectors than UTP connectors, is more expensive than UTP, and requires careful grounding to work properly.

Getting wired

If you're wiring a home or a small office, you can usually do it yourself. For large cable installation jobs or for running cable through walls, you might want to call in a professional.

Category 5 twisted-pair cable is readily available in retail computer stores and mail-order catalogs. You can buy it in a wide variety of lengths, such as 10, 25, and 50 feet, with male RJ-45 connectors at each end. Cable sold in specific lengths with connectors at each end are often labeled *patch cables*.

The connectors at each end of the TP cable are RJ-45 male connectors that insert into female jacks in the NIC and the hub or switch port. You can use Category 5 data couplers to connect two cables and extend the cable length. Many UTP cable vendors now offer color-coded cables that you can use to help identify which cable is connected to which device. You need to get a single 10BaseT cable with an RJ-45 connector on each end.

A number of helpful cabling solutions can help you run 10BaseT cabling in your home or business. Adhesive-backed, vinyl raceways allow you to run 10BaseT cabling along walls, floorboards, and floors (hiding the cables and keeping people from tripping on them.

 Cable run through walls is terminated at a wall plate, much the way a telephone jack is installed. You need to use a fire-rated cable, which is more expensive than regular cabling. If you want your network cabling installed inside walls or over drop ceilings, you might want to call in a cable contractor. Make sure you use someone with computer network cabling experience. Although electricians can lay the cable, they often lack the necessary network skills. They also typically lack the right test equipment to verify the integrity of the cable.

One-on-one with a crossover cable

If you're connecting a DSL modem or router directly to a single computer, you might need to use a special type of twisted-pair cable called a crossover cable. In a *crossover cable,* one of the four pairs is crossed to allow a LAN device to be connected directly to a single NIC instead of to the hub.

Cross-wired cables are usually designated with a color patch that differentiates them from other cables. If the crossover cable isn't marked, you should mark it so that it doesn't become mixed up with standard TP cables. You can check the connectors at each end to determine whether a TP cable is a crossover cable. If one of the wire pairs is crossed over at the connector, it's a crossover cable. Some DSL bridges and routers include special RJ-45 ports for connecting to a single computer using a typical fire-rated cable instead of a crossover cable.

Phoneline cabling

Phoneline networking uses standard telephone wiring with RJ-11 connectors that are like the ones you use to connect telephones, modems, and faxes to the telephone line. Unlike 10BaseT cabling with four-wire pairs, telephone wire consists of two pairs of copper wires that are twisted around each other to cancel out interference.

The cabling layout is different for a Phoneline network than for an Ethernet network. There is no star topology. You can connect individual computers with Phoneline NICS to telephone wall jacks in different rooms.

You can also connect more than one PC in a room with only one telephone jack by using a daisy chain topology. You create this daisy chain by linking the computers in the room to the computer connected to the telephone wall jack. You connect one Phoneline adapter to another using telephone wiring connected to the available RJ-11 jack on the Phoneline NICs.

 If you have only one phone jack in your room and you want to use your modem or fax, purchase a two-way adapter, which turns your single phone jack into a double phone jack. Plug your telephone or fax into one phone jack, and plug your Phoneline NIC into the other.

Journey to the Center of the LAN

At the center of a star topology Ethernet network is a hub or a switch that keeps the data traffic moving to computers and other network devices on your network. On a wireless LAN that uses a star topology, an Access Point is the wireless equivalent to the hub.

Ethernet hubs and switches

A hub or a switch is at the center of an Ethernet LAN, with all computers — including your Ethernet DSL modem or router — connecting to it through cables. The fundamental difference between hubs and switches is one of intelligence. Hubs are passive devices that repeat all traffic to all ports. Switches can actively segment a network and forward packets to only the ports that need them, dramatically lowering traffic levels and increasing network performance. Hubs and switches can be connected in different configurations to better manage a network or to expand a LAN.

Most Fast Ethernet hubs and switches support both Ethernet and Fast Ethernet. These devices are often referred to as *10/100 hubs,* or *switches.* Each network device connects to a hub or a switch through a port that's a female RJ-45 connector. Hubs and switches come in many port densities; port counts of 8, 12, 16, and 24 are the most common.

Stackable hubs and switches allow an Ethernet LAN to start out small and grow as needed because you can literally stack them on top of each other to form a single network with many hubs. Most hubs include uplink ports to connect one hub to another using a straight-through Ethernet cable instead of a crossover one. The maximum distance you can have between two hubs that are using Fast Ethernet is about 16 to 32 feet (5 to 10 meters).

Fast Ethernet allows only two hubs to be connected. Figure 11-11 shows how a hub forming the basis of one network can be linked to another hub to create a single, larger network.

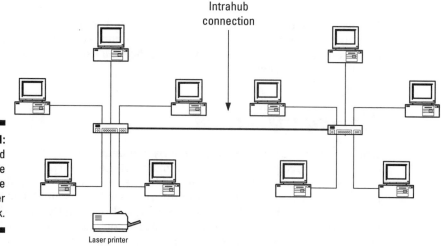

Figure 11-11: Two linked hubs create a single larger network.

What's all the hub-bub?

Hubs are simple, passive devices that keep network data traffic moving. Hubs designed for small networks typically include four to eight ports and are about the size of a large external modem. Hubs have LEDs on the front to tell you what's going on with network traffic and your node connections. Figure 11-12 shows a 3Com 10/100 hub.

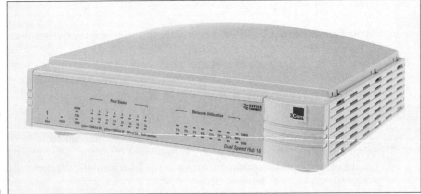

Figure 11-12: The 3Com Dual Speed Hub 16.

Photo courtesy 3Com Corporation

Hubs that support Ethernet and Fast Ethernet are called autosensing, or dual-speed, hubs. These hubs adjust the ports to either 10 Mbps or 100 Mbps, depending on the NIC connected to the port. Autosensing hubs cost between $100 to $300.

Intelligent hubs, or *managed hubs,* may also include network management software that lets you monitor hub status from a remote console. Typically, homes and small businesses with a single location use unmanaged hubs, which are less expensive.

Making the switch

When a network uses a hub, data sent by workstations is passed around the entire network, regardless of the destination of the data. This results in a lot of unnecessary traffic bouncing around the network, which can reduce the LAN's performance. A switch solves this problem because it listens to the network and automatically learns what workstations can be reached through what ports. The switch then selectively passes on data by transmitting the traffic from only the relevant port instead of all the ports, the way a hub does.

Switching technology has a big effect on network performance. Switches segment a network and provide extra bandwidth by managing data traffic. Switches are self-learning, so they learn about your network and keep up with any network configuration changes.

Switches are more expensive than hubs, but you can get a small 4-port switch for under $100. Switches designed for small networks come in a variety of port configurations. Most switches are dual speed, 10/100 switches. Like hubs, switches are usually stackable with other hubs and switches from the same vendor. Figure 11-13 shows a 3Com switch.

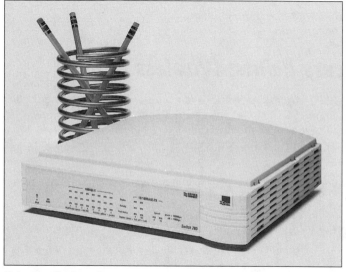

Figure 11-13: A switch from the 3Com Office Connect family of networking properties.

Photo courtesy 3Com Corporation

You can connect hubs to a switch to increase the number of workstations connected to your network. These hubs can be either Ethernet or Fast Ethernet hubs. By connecting your existing hubs to each other, you can leverage your investment and harness the power of switching. A switch can improve the performance of your DSL connection by isolating that segment of your network from unnecessary data traffic.

Use a switch with Fast Ethernet for one or more of the following reasons:

✔ **To overcome Fast Ethernet topology limitations.** Fast Ethernet has a diameter limitation of 200 meters. Switches can be used to overcome topology limitations by breaking a single segment into multiple segments.

✔ **To improve performance by segmenting a Fast Ethernet network.** When a single segment gets too busy, users complain about response times, connections drop, or access to the Internet slows down. As the number of users on a segment increases or their demands on the network grow, network utilization drops. Using a switch can solve this problem.

✔ **To provide high network performance for servers or network devices.** Any single computer, such as a network server or a DSL modem or router, can be connected to a switch that supports full duplex. Full-duplex operation allows information to be transmitted and received simultaneously and, in effect, doubles the potential throughput of the link. When you attach a switch with a full-duplex link to a server, you won't have any collisions when transmitting. This allows the network segment to be more efficient.

Access Points: Wireless hubs

An 802.11b wireless network uses a wireless hub-like device called an Access Point. Figure 11-14 shows the Cisco Aironet 340 Series Wireless Bridge. The Access Point acts as the hub for all computers with wireless NICs installed and acts as a bridge for connecting to an Ethernet network.

Figure 11-14: The Cisco Aironet 340 Series Wireless Bridge acts as a hub for a wireless LAN and a bridge to an Ethernet network.

Photo courtesy Cisco

The wireless-to-Ethernet bridge function is important because it allows the wireless LAN to connect to the Ethernet LAN to access network services, such as printing and Internet access through the DSL modem.

Macs Are Made for Networking

Apple has fully integrated networking into its latest Mac product line. Ethernet, Phoneline, and wireless networking are all supported on the Macintosh.

Mac NICs

iMacs, iBooks, PowerBooks, and Power Mac G4 have built-in 10/100 Ethernet adapters, and most support Apple's AirPort 11-Mbps wireless networking. Because the Mac as well as the PC uses Ethernet, you can connect Macs to PCs using Ethernet. You connect your Macs on an Ethernet LAN in the same way you do for PCs, except you don't need to install a NIC.

If you plan to use a Mac as the gateway for your DSL connection, you need to install a second NIC in the Power Mac G4 using one of the PCI slots or you can install a USB NIC. Farallon (www.farallon.com) offers a USB NIC for the Mac.

Wireless Mac networking with AirPort

Apple's wireless network solution, called AirPort, can be added to Macs to support 11-Mbps wireless networking. The Power Mac G4, iMac, iBook, and Powerbook support AirPort wireless networking through slots for inserting the AirPort adapter card.

The AirPort Card retails for $99 and the AirPort Base Station retails for $299. The AirPort Base Station is the wireless network Access Point; it has a cool, flying saucer look, as shown in Figure 11-15. In addition to supporting the wireless network, the AirPort Base Station contains a 56K modem and a 10BaseT Ethernet port for connecting to a DSL modem or LAN. AirPort is based on the IEEE 802.11b standard. The range of the AirPort wireless network is typically around 150 feet, even through walls.

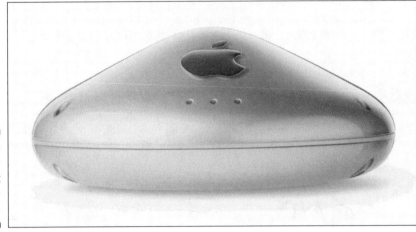

Figure 11-15:
The Apple
AirPort
Base
Station.

Photo courtesy Apple Computer Corporation

You can also use a software version of the AirPort Base Station, which allows an AirPort-enabled computer to act as the wireless base station through its Internet connection. With AirPort Software Base Station, you can connect a wireless LAN to a computer using a PCI or USB DSL modem.

Chapter 12

Windows 95/98 Meets DSL

*I*f you want to use a computer running Windows 95/98 to access the Internet through DSL, you need to configure Microsoft TCP/IP. This chapter explains the fundamentals of configuring Windows 95/98 TCP/IP for connecting to the Internet through your DSL CPE.

If your Windows 95/98 computer is configured as part of your DSL service, you can skip this chapter.

Windows 95/98 and DSL Connections

You can use DSL CPE that connects to your PC through an Ethernet, a PCI, or an USB interface. (For more on this topic, read Chapter 8.) The way that the DSL CPE connects to your computer defines how you configure the TCP/IP settings for your Internet connection.

Depending on the DSL Internet service package you choose, the DSL provider might configure your Windows 95/98 computer for TCP/IP as part of the service installation or provide installation software that automatically configures the TCP/IP settings for you. If your Windows 95/98 computer is configured as part of your DSL service, you can skip this chapter.

Here is how different DSL CPE interface with Windows 95/98 and where you configure their TCP/IP settings:

✔ If you use **an Ethernet DSL modem or router,** Windows 95/98 uses the Ethernet interface through a NIC (network interface card). Using an Ethernet solution means that you configure the NIC card for TCP/IP.

✔ If you use **a DSL USB modem that doesn't use PPP over DSL,** Windows 98 uses the Ethernet interface. You configure the DSL CPE device driver for TCP/IP as you would configure a NIC.

✔ If you use **a PCI modem that doesn't use PPP over DSL,** Windows 95/98 uses the Ethernet interface. You configure the DSL CPE device driver for TCP/IP as you would configure a NIC.

✔ If you use **a DSL USB modem that uses PPP over DSL,** the modem is installed as a NIC but Windows 98 uses Dial-Up Networking (DUN) to make the Internet connection through DSL.

✔ If you use **a DSL PCI modem that uses PPP over DSL,** the modem is installed as a NIC but Windows 95/98 uses Dial-Up Networking (DUN) to make the Internet connection through DSL.

Windows 98 SE (Second Edition) is an enhanced version of Windows 98 that includes Windows 98 bug fixes. One of the most important fixes is improved USB support. If you plan on using a USB DSL modem, upgrade to Windows 98 SE before you install the modem.

Setting Up TCP/IP in Windows 95/98

The first step in configuring Windows 95/98 is to make sure the Microsoft TCP/IP stack is installed. If you've been connecting to the Internet, the Microsoft TCP/IP stack is already installed.

You can check to see whether the TCP/IP stack is installed by right-clicking the Network Neighborhood icon and choosing Properties from the pop-up menu. The Network properties dialog box appears, as shown in Figure 12-1.

In the Configuration tab, you should see the following entry:

```
TCP/IP ⇨ network card adapter
```

where *network card adapter* is the name of the installed network adapter card. If you see an entry like the following in Network properties, the Microsoft TCP/IP stack is already installed:

```
TCP/IP ⇨_3Com Fast EtherLink XL 10/100Mb TX Ethernet NIC
        (3C905B-TX)
```

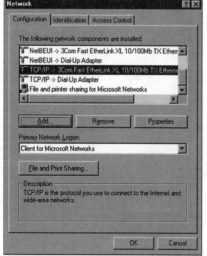

Figure 12-1:
If TCP/IP is installed, it appears here, bound to a NIC or a dial-up adapter.

If you don't have a NIC installed, TCP/IP is bound to the Dial-Up Adapter in the Network properties dialog box. The entry appears as the following (see Figure 12-1):

```
TCP/IP ⇨ Dial-Up Adapter
```

If you don't have TCP/IP installed, you need to install the Microsoft TCP/IP stack. Doing so binds TCP/IP to any installed NIC or DSL modem driver in the Network properties dialog box. Binding is where a given networking protocol software is attached to the physical network adapter.

Before installing the Microsoft TCP/IP stack, have the original Windows 95/98 distribution media handy, in case Windows prompts you for it.

To install the Microsoft TCP/IP stack:

1. **Right-click the Network Neighborhood icon on your desktop and then choose Properties from the pop-up menu.**

   Alternatively, you can double-click the Network icon in the Control Panel. The Network properties dialog box appears.

2. **In the Configuration tab, click the Add button.**

   The Select Network Component Type dialog box appears.

3. **Double-click Protocol.**

   The Select Network Protocol dialog box appears, as shown in Figure 12-2.

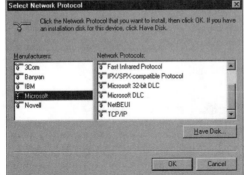

Figure 12-2:
The Select
Network
Protocol
dialog box.

4. **In the Manufacturers list on the left side of the dialog box, select Microsoft.**

The available protocols appear in the Network Protocols list on the right side of the dialog box.

5. **Double-click the TCP/IP item.**

When you return to the Network dialog box, scroll through the list of network components and you see TCP/IP listed. Notice that the component entry for TCP/IP uses the following notation:

```
TCP/IP ⇨ network adapter name
```

TCP/IP is bound to your network adapter card.

6. **Follow the Windows prompts to complete the installation.**

Windows 95/98 might prompt you for the original distribution CD-ROM or disks to complete the installation.

Configuring Your NICs for TCP/IP

The next step towards DSL enlightenment is to configure your Windows 95/98 PCs for the type of IP network configuration you're using for your DSL connection. This process involves configuring the TCP/IP stack, which is bound to your NIC.

You can configure your TCP/IP stack in two ways:

✔ If you're using a DSL router or any other Internet-sharing solution that uses NAT, you'll be using the router's DHCP (Dynamic Host Configuration Protocol) server to dynamically assign IP addressing information to each computer on your network. You configure your Windows 95/98 clients as DHCP clients and don't add any specific IP address information to the TCP/IP properties.

✔ If you're using static IP addresses assigned to specific computers, manually configure the TCP/IP properties of each Windows 95/98 PC with IP address and DNS information.

Windows 95 and 98 allow only one TCP/IP stack configuration profile for LAN-to-Internet (or other network) connections. This means if you have more than one ISP connection available, you must change the settings in the TCP/IP Properties dialog box for each connection and then reboot the system. A shareware program called NetSwitcher (www.netswitcher.com) enables you to create multiple TCP/IP network connections and switch between them as needed. For example, if you're using a laptop and travel between different offices, each with its own LAN-based Internet access service, you can use NetSwitcher to create profiles for all your different TCP/IP network connections and switch to the one you need when you're at a specific office. NetSwitcher also makes it easy to switch between a dedicated DSL connection and a dial-up connection when you're on the road.

USB and PCI DSL modems are NICs in Windows 95 and 98

When you install a USB or PCI DSL modem in your Windows 98 computer or install a PCI DSL modem in Windows 95, the device is installed as a network adapter. From your PC's perspective, the USB or PCI DSL modem appears as a network interface card (NIC).

If your DSL service uses a PCI or USB modem without PPP over DSL, you configure the TCP/IP settings for your computer using the Network properties dialog box in Windows 95 or 98. You select the DSL modem driver in the Network properties dialog box and then click the Properties button.

Configuring for dynamic IP addressing

Using a DSL router or other Internet-connection-sharing solution that acts as a DHCP server enables dynamic IP addressing. This type of IP addressing is commonly used for consumer DSL offerings. You might get your IP address dynamically assigned by the ISP for your DSL connection. Or you might get only one IP address, which means you need to use a router or some other Internet-sharing solution to share the DSL connection across multiple computers.

The DHCP server assigns an IP address from an available pool of IP addresses for use only during the current connection. The DHCP server functionality of the router allows IP addresses, subnet masks, and default gateway addresses to be assigned to computers on your LAN. You can use DHCP with registered IP addresses or private IP addresses along with the NAT (network address translation) feature incorporated in most DSL routers. Chapter 7 explains IP addressing and DHCP in more detail.

Windows 95 and 98's default TCP/IP stack configuration is as a DHCP client. A DHCP server will detect your PCs acting as DHCP clients. You need to config- ure your DSL router or Internet-connection-sharing solution to act as a DHCP server and tell it what IP addresses you plan to use.

If you've installed TCP/IP for the first time, you don't need to change any default settings to support DHCP. If you're unsure of your TCP/IP configura- tion, however, you can double-check to make sure the default TCP/IP settings are in place by performing the following steps:

1. **Right-click the Network Neighborhood icon on the desktop and then choose Properties from the pop-up menu.**

   The Network properties dialog box appears.

2. **In the network components list, select the network adapter with the TCP/IP binding and then click the Properties button.**

   The TCP/IP Properties dialog box appears.

3. **Click the IP Address tab (shown in Figure 12-3) and select the Obtain an IP address automatically option (the default setting).**

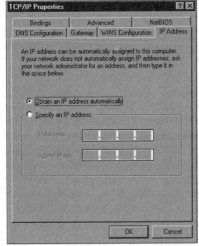

Figure 12-3:
The IP
Address
tab in the
TCP/IP
Properties
dialog box.

4. **Click OK twice to exit the Network applet.**

5. **Restart Windows.**

After you configure your Windows clients for DHCP, whenever you boot the Windows 95/98 computer, it will get the IP address information it needs from the DHCP server.

Configuring for static IP addressing

If you plan on using registered IP addresses that you want to assign to specific computers — to run a Web server, for example — you need to configure each Windows 95/98 workstation with static, routable IP address and DNS information. You can also use private (unregistered) IP addresses, but in most cases, you use private IP addresses with NAT and DHCP.

Before you configure your Windows 95/98 clients for static IP addressing, you need to get the following information from your ISP:

- An IP address for each Windows 95/98 computer that you're connecting to the Internet

- A subnet mask IP address for your network

- The gateway IP address, which is the IP address assigned to your DSL router or your ISP's router if you're using a DSL modem

- A host and domain name for the IP address

- DNS server IP addresses

Follow these steps to configure a Windows 95/98 workstation for static IP addressing:

1. **Right-click the Network Neighborhood icon on the desktop and then choose Properties from the pop-up menu.**

   The Network properties dialog box appears.

2. **Double-click TCP/IP ➪ *network adapter name* in the network components list, or select it and then click the Properties button.**

   The *network adapter name* is whatever the name of your network adapter as it appears in the components list.

   The TCP/IP Properties dialog box appears.

3. **Click the IP Address tab**.

4. **Select the Specify an IP address option**.

   The IP Address and Subnet Mask fields become active.

5. **In the IP Address field, enter the IP address for the Windows 95/98 workstation you're configuring. In the Subnet Mask field, enter the subnet mask IP address.**

6. **Click the DNS Configuration tab and then select the Enable DNS option.**

   All the controls in the Enable DNS group become active, as shown in Figure 12-4.

7. **In the Host field, enter the host name for the Windows 95/98 workstation uniquely associated with the specific IP address you entered in the IP Address tab.**

   The host machine might be named david, which added to angell.com becomes david.angell.com.

8. **In the Domain field, enter your organization's domain name.**

   For example, the host machine might be named david and the domain name might be angell.com.

9. **In the DNS Server Search Order area, enter the first DNS server IP address that you want your Windows 95/98 workstation to check, and then click the Add button. Do the same for each additional DNS server address you have.**

   To remove any DNS server IP address from the list, select it and then click the Remove button.

10. **Click the Gateway tab, which is shown in Figure 12-5.**

11. **In the New gateway field, enter the IP address of your DSL router or the ISP's router if you're using a DSL modem (bridge) and then click the Add button.**

    If you have access to multiple DSL connections, you can add multiple gateway IP addresses. Windows 95/98 will go through the list in sequential order until it finds the gateway address that works. To remove any gateway entry from the list, select it and click the Remove button.

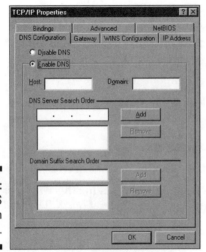

Figure 12-4:
The DNS
Configuration
tab.

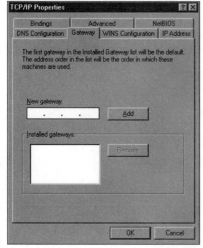

Figure 12-5:
The
Gateway
tab.

12. **Click OK twice to exit the TCP/IP Properties and Network dialog boxes.**

13. **If prompted, insert your distribution CD-ROM.**

    After installing the necessary information, Windows asks you to restart your computer.

14. **When prompted, click Yes to restart Windows with the new TCP/IP settings.**

Managing protocol bindings

Windows 95/98 supports multiple network adapters, and each of these cards can have different network protocols bound to it. For a given protocol to communicate with each network adapter on your computer, the network adapter driver must be bound to the protocol. *Binding* a protocol to a network adapter means you're attaching a software network interface to the network adapter. The binding defines the relationships between networking software components.

Whenever you install a new adapter card, Windows 95/98 automatically binds some protocols to it. You probably want to change the bindings — typically, you must add or remove bindings as your network changes.

You can change network adapter protocols by selecting the Bindings tab in the Properties dialog box of the selected network adapter. To display that dialog box, select the driver for the network adapter (the item in the list without a protocol name in front of it) in the Network properties dialog box, and then click the Properties button. Figure 12-6 shows the network adapter Properties dialog box with two network protocols available, NetBEUI and TCP/IP. Note that only TCP/IP is bound to the NIC.

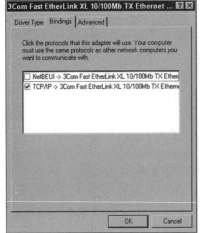

Figure 12-6:
The network
adapter
Properties
dialog box.

The general rule for binding protocols is to bind only the protocols that you need. Leaving unnecessary protocols bound to your network adapter slows down network performance and can cause other problems. If you're not using it, remove it. You can add a removed protocol at any time.

Working the DUN Way

If you have a PCI or USB DSL modem that uses PPP over DSL, you need to use Windows 95/98 Dial-Up Networking (DUN). If you don't have any DSL CPE that uses PPP over DSL, you don't have to do anything with Dial-Up Networking.

Depending on the your DSL service, your Windows 95/98 computer might be configured for DUN and TCP/IP as part of the service installation. Or the installation software provided with your service might have automatically configured the DUN and TCP/IP settings for you.

The DUN connection profile provides Windows 95/98 with the information it needs to make the DSL connection to the Internet. The creation of a DUN connection profile mimics the creation of a DUN profile for an analog or ISDN modem — with one major difference. The entry for the telephone number is a

Virtual Circuit (or Channel) Identifier (VCI) number. This 16-bit number is provided by the DSL ISP and identifies the channel your DSL service will use to make the connection.

If you're using Windows 95, you need to use DUN version 1.2 or higher. Some DSL modem vendors include the latest version of DUN as part of their installation software. You can download the DUN upgrade from the Microsoft Web site (www.microsoft.com/msdownload/). If you're using Windows 98, you can use the DUN software that comes with Windows 98.

Installing Dial-Up Networking

You can check to see whether DUN is installed by opening the My Computer icon on your desktop. If you see an icon labeled Dial-Up Networking, DUN is installed. If you don't see that icon, you must install the Dial-Up Networking. Before you can install DUN, however, you need to make sure that the DSL modem adapter card or USB modem is installed.

Before you proceed to install Dial-Up Networking, make sure you have your original Windows 95/98 distribution disks or CD, in case you're prompted for them.

Here's how to install Dial-Up Networking:

1. **In the Control Panel, double-click the Add/Remove Programs icon.**

   The Add/Remove Programs Properties dialog box appears.

2. **Click the Windows Setup tab.**

3. **Double-click the Communications option.**

   The Communications dialog box appears, displaying the available communications and networking components of Windows 95/98.

4. **Select Dial-Up Networking (see Figure 12-7) and then click OK.**

   This returns you to the Windows Setup page.

5. **Click OK.**

6. **Follow the Windows 95/98 prompts for installing the disks or CD-ROM that Windows needs to install Dial-Up Networking.**

7. **When prompted, restart Windows 95/98.**

Figure 12-7: The Communications dialog box displays which Windows 95/98 components are installed and which aren't.

Setting up a connection profile

If you installed PPP over DSL CPE, a DUN profile might have been created for you. Therefore, you should check your Dial-Up Networking window before you create a Dial-Up Networking profile.

To create a DUN connection, do the following:

1. **Double-click the My Computer icon on the desktop, double-click the Dial-Up Networking folder, and then double-click the Make New Connection icon.**

   The Make New Connection wizard appears.

2. **Click Next.**

3. **Type a name (such as** DSL**) to identify your connection and then select your DSL device from the list of modems.**

   This list includes any USB modems as well as any modem cards that you've installed.

4. **Click Next.**

   A page appears for you to type the telephone number of the host computer that you want to call.

5. **Enter any telephone number or type the Virtual Circuit Identifier number, and then click Next.**

   Although DUN doesn't use the telephone number to make your DSL connection, you can't complete the connection profile unless you enter a number.

6. **Click Finish.**

Configuring your DUN connection profile

After you create a DUN connection profile, you might need to configure it for TCP/IP networking so that your computer can connect to the Internet through the DSL modem.

Does your DSL service uses static IP addresses? If so, before you configure your DUN profile Windows 95/98, you need the IP address for your computer and the DNS server IP addresses from the ISP.

You use the Server Types dialog box to configure TCP/IP for the DUN profile. This dialog box includes a list of protocols for dial-up connections. You configure the TCP/IP protocol for Internet connections as follows:

1. **Double-click the My Computer icon on the desktop and then double-click the Dial-Up Networking icon.**

   You see a new icon with the name you entered for your connection.

2. **Right-click the connection icon and choose Properties from the pop-up menu.**

   A connection profile dialog box appears with the name of your connection in the title bar.

3. **Click the Server Types tab, which is shown in Figure 12-8.**

4. **In the Type of Dial-Up Server list, Windows 95 users should select** *PPP: Windows 95, Windows NT 3.5, Internet.* **Windows 98 users should select** *PPP: Internet, Windows NT Server, Windows 98.*

5. **In the Allowed network protocols area, click to add a check mark to the TCP/IP item. Remove any other check marks in that area (by clicking).**

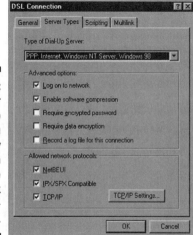

Figure 12-8:
The Server Types tab lets you specify information about the network you're connecting to.

6. **Click the TCP/IP Settings button.**

The TCP/IP Settings dialog box appears, as shown in Figure 12-9.

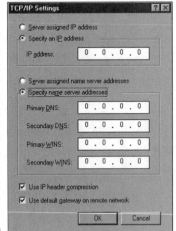

Figure 12-9:
Use the
TCP/IP
Settings
dialog box
to enter
the IP
addresses
of the
network
you're con-
necting to.

7. **Depending on the type of IP addressing used for your Internet access, do one of the following:**

   • If your ISP uses dynamic IP addressing, select the Server assigned IP address option at the top.

   • If your service provider has assigned you a specific IP address, select Specify an IP address. Enter the IP addresses assigned to your computer by the ISP and then enter the IP address in the Primary DNS and Secondary DNS boxes.

8. **Click OK three times.**

Making the DUN connection

The easiest way to make a DUN connection is to double-click any TCP/IP application, such as the Internet Explorer or Netscape Communicator icon on your desktop. The browser window opens along with the Connect To dialog box. Click the Connect button. The connection is made and the home page appears in your Web browser. To disconnect, click the minimized Connect To dialog box on the taskbar and then click Disconnect.

You can access the DUN connection profile for your DSL connection also by opening the Dial-Up Networking folder in My Computer to display the Dial-Up Networking window (see Figure 12-10). Double-click the connection profile for your DSL connection. The Connect To dialog box appears (see Figure 12-11).

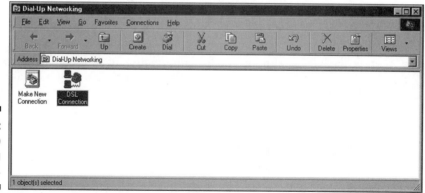

Figure 12-10:
The Dial-Up
Networking
window.

Figure 12-11:
The Connect
To dialog
box.

You need to enter your username and password in the appropriate fields. The Save password check box below the Password field, if checked, will tell Dial-Up Networking to remember the password so that you won't have to enter it on subsequent dial-up sessions. For a DSL connection, the number in the Phone number field is an encoded number or a Virtual Circuit Identifier (VCI) number assigned by your ISP.

One of the big advantages of a DSL connection is that it is always on, so you will not be connecting and disconnecting as much as you did with a dial-up Internet connection. If you want to break a connection, however, display the Dial-Up Networking status box by clicking the DUN icon on the right side of the Windows 95/98 taskbar. A Status dialog box appears; this dialog box varies depending on the DSL modem used. To terminate your session, click the Close button. The link terminates immediately.

Bypassing the Connect To dialog box

Every time you use a Dial-Up Networking connection, the Connect To dialog box appears, prompting you for confirmation of your connection before making the call. You can get rid of this Connect To dialog box and automatically make the dial-up connection by unchecking the Prompt for information before dialing setting in the Dial-Up Networking dialog box (see Figure 12-12).

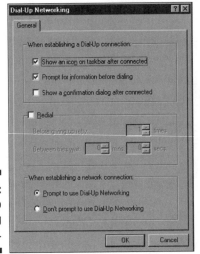

Figure 12-12:
The Dial-Up
Networking
dialog box.

Two Handy TCP/IP Utilities

Windows 95/98 includes two useful TCP/IP utilities called winipcfg and ping that help you manage your TCP/IP connection. This section explains how to use these two TCP/IP utilities.

What's your IP status?

The winipcfg program is a TCP/IP utility that lets you view information about your TCP/IP protocol and network adapter settings. If you're using DHCP on your LAN to receive an IP address from a host, the winipcfg program is a handy way to figure out the IP address that's been assigned to your PC and how long it's been assigned. And the winipcfg program lets you reset your IP address settings when Windows 95/98 doesn't properly get the dynamic IP address information from the DHCP server when you boot up your computer.

The winipcfg program (winipcfg.exe) is located in your Windows folder. To run the winipcfg program, click Start ➪ Run. In the Open box, type **winipcfg.exe** and then click OK. The IP Configuration dialog box appears, as shown in Figure 12-13. Here, you can see the physical address, IP address, subnet mask, and default gateway settings for your primary TCP/IP network adapter card.

Figure 12-13: The IP Configuration dialog box.

Click the More Info button to display an expanded IP Configuration dialog box, as shown in Figure 12-14. This dialog box displays such information as your computer's host name, the address of the DHCP server (if you're using one), and DNS server IP addresses.

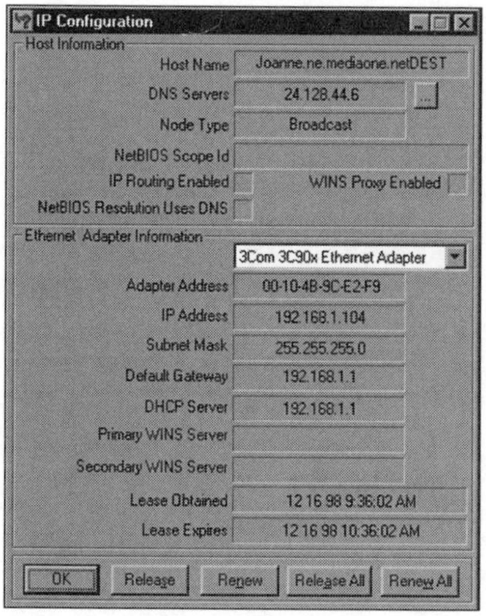

Figure 12-14: The expanded IP Configuration dialog box.

Information in the expanded IP Configuration dialog box is broken down into two groups: Host Information and Ethernet Adapter Information. Also included is information on a variety of IP and DNS settings and network adapter information (such as the hardware address). The Release and Renew buttons at the bottom of the dialog box enable you to release and renew the assigned IP address, respectively.

You can print the information in the IP Configuration dialog box by first copying the data to Word or NotePad. To do this, click the Winipcfg control menu, which is the small icon in the upper-left corner of the IP Configuration window title bar. Choose Copy or press Ctrl+C. The contents of the Winipcfg window are copied into the Windows Clipboard, enabling you to paste the information into another application. After the information is in the application (such as Word or Notepad), you can manipulate the text or print it.

The winipcfg program is such a handy program to have accessible that you should have a shortcut to it on your Windows desktop. The easiest way to do this is to click Start ➪ Find ➪ Files or Folders, enter **winipcfg.exe** in the Named field, select the drive containing the Windows directory, and then click Find Now. Drag the winipcfg file to the desktop. A shortcut to winipcfg now appears on your desktop.

Ping, ping, ping

The ping program is a handy TCP/IP diagnostic utility that works like TCP/IP sonar: You send a packet to a remote host and, if the host is functioning properly, it bounces the packet back to you. Ping prints the results of each packet transmission on your screen. By default, Ping sends four packets, but you can use it to transmit any number of packets or transmit continuously until you terminate the command.

The ping program is the single most useful program for troubleshooting the status of any TCP/IP connection. The ping command includes a number of parameters, which you can view by typing **ping** (at the DOS prompt) and pressing Enter.

You can access ping from the MS-DOS Prompt window (Start ➪ Programs ➪ MS-DOS Prompt). At the DOS prompt, enter the command in the following format:

```
ping IPaddress
```

where *IPaddress* is the IP address of the remote host you want to check. For example, if you type the following:

```
ping 199.232.255.113
```

and press Enter, you ping the host machine with the IP address of 199.232.255.113. The host machine can be on your local network or on the Internet. Likewise, you can ping the host name instead of the IP address for sites for which you don't know the IP address. For example, entering the following:

```
ping www.angell.com
```

returns the same ping information that you would get using the IP address. Figure 12-15 shows a sample ping result as it appears in the MS-DOS Prompt window.

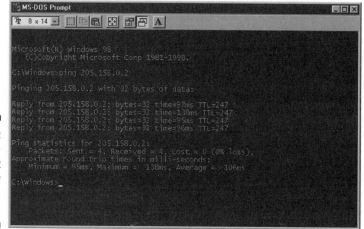

Figure 12-15: The MS-DOS Prompt window with ping results.

If the packets don't come back after you execute the ping command, either the host is not available or something is wrong with the connection.

Chapter 13

Windows NT Meets DSL

● ●

In This Chapter

▶ Connecting DSL to Windows NT

▶ Finding out how to set up TCP/IP in Windows NT

▶ Discovering how to configure NICs for TCP/IP

▶ Installing and configuring Windows NT Remote Access Service

▶ Using Dial-Up Networking for DSL connections

▶ Working with handy Windows NT TCP/IP utilities

● ●

*T*o use a computer running Windows NT to access the Internet using DSL, you need to configure the Microsoft TCP/IP. This chapter explains the fundamentals of configuring Windows NT TCP/IP for connecting to the Internet through your DSL CPE.

If your Windows NT computer is configured as part of your DSL service, you can skip this chapter.

Windows NT and DSL Connections

You can use DSL CPE that connect to your PC using the Ethernet, PCI, or USB interface. (For more about this, see Chapter 8.) The way the DSL CPE connects to your computer defines how you configure the TCP/IP settings for your Internet connection.

Depending on the DSL Internet service package you choose, the DSL provider might configure your Windows NT computer for TCP/IP as part of the service installation or provide installation software that automatically configures your TCP/IP settings for you. If your Windows NT computer is configured as part of your DSL service, you can skip this chapter.

Windows NT does not include native USB support.

Here is how different DSL CPE interface with Windows NT and where you configure their TCP/IP settings.

✔ If you use **an Ethernet DSL modem or router,** Windows NT uses the Ethernet interface through a NIC (network interface card). Using an Ethernet solution means that you configure the NIC card for TCP/IP.

✔ If you use **a DSL PCI modem that doesn't use PPP over DSL,** Windows NT uses the Ethernet interface. You configure the TCP/IP for the DSL CPE device driver as you would configure a NIC.

✔ If you use **a DSL PCI modem that uses PPP over DSL,** the modem is installed as a NIC but Windows NT uses Dial-Up Networking (DUN) to make the Internet connection through DSL.

Setting Up TCP/IP in Windows NT

Setting up Windows NT for a DSL connection involves installing and configuring the Microsoft TCP/IP stack, which provides the TCP/IP networking protocol for connecting to the Internet. When you install the TCP/IP stack, you also bind the TCP/IP protocol to your network adapter card and Dial-Up Networking (if it's installed), which is part of Windows NT's Remote Access Service. *Binding* is a process that defines the relationship between a network adapter card and network protocols used with that network card. When you bind the TCP/IP protocol to a network adapter, you allow TCP/IP traffic to pass through your Ethernet network to the DSL modem or router. Likewise, binding TCP/IP to the Dial-Up Networking adapter enables you to use TCP/IP for DSL connections using Dial-Up Networking.

The first step in configuring Windows NT is to install the Microsoft TCP/IP stack, if it's not already installed. If you've been connecting to the Internet, Microsoft TCP/IP is already installed.

You can check to see whether the TCP/IP stack is installed by right-clicking Network Neighborhood, choosing Properties from the pop-up menu, and then choosing the Protocols tab (see Figure 13-1). If you see TCP/IP Protocol in the Network Protocols list, the Microsoft TCP/IP stack is already installed. You can skip the following steps, but you still need to make sure that TCP/IP is bound to the NIC, which is explained later in this chapter.

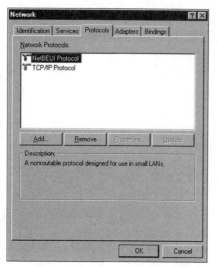

Figure 13-1:
The
Protocols
tab in the
Network
properties
dialog box
shows the
installed
networking
protocols.

If you don't have TCP/IP installed, you need to install the Microsoft TCP/IP stack. Installing the Microsoft TCP/IP stack binds TCP/IP to any installed NIC or DSL modem driver in the Network properties dialog box. Before installing the Microsoft TCP/IP stack, have the original Windows NT distribution media handy, in case Windows prompts you for it.

To install the Microsoft TCP/IP stack, do the following:

1. **Right-click the Network Neighborhood icon on the desktop and choose Properties from the pop-up menu.**

   Alternatively, you can double-click the Network icon in the Control Panel. The Network properties dialog box appears.

2. **Click the Protocols tab to display the Protocols properties.**

3. **Click Add.**

   The Select Network Protocol dialog box appears, as shown in Figure 13-2.

4. **Select the TCP/IP Protocol and then click OK.**

5. **When Windows NT asks whether you want to use the Dynamic Host Configuration Protocol (DHCP) when using TCP/IP, click No.**

   If you add DHCP server functionality to your LAN, you can always change the TCP/IP settings, as explained later in this chapter.

6. **If you have Remote Access Service (RAS) installed, click Yes when Windows NT asks whether you want to configure it to use TCP/IP.**

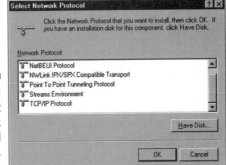

Figure 13-2:
The Select
Network
Protocol
dialog box.

7. **When Windows NT displays the Windows NT Setup dialog box to prompt you for the location of the distribution files, enter the drive and directory where the Windows NT 4.0 distribution files are located.**

8. **Click Continue.**

   After the files are copied, Windows NT displays the Network dialog box with the TCP/IP Protocol visible.

9. **Click Close to complete the installation.**

   After the bindings have been configured, Windows NT displays the Network Settings Change dialog box to notify you that you must restart your computer before the changes will take effect.

10. **Click Yes to restart the computer.**

Configuring Your NICs for TCP/IP

When you bind the TCP/IP protocol to a network adapter, you allow TCP/IP traffic to pass through your Ethernet network to the DSL bridge or router. You need to configure the Microsoft TCP/IP stack bound to the NIC for the IP address configuration you're using on your LAN.

You can configure your TCP/IP stack in two ways:

✔ If you're using a DSL router as a DHCP (Dynamic Host Configuration Protocol) server to dynamically assign IP addresses, configure your Windows NT computers as DHCP clients, and don't add specific IP address information to the TCP/IP properties.

> ✔ If you're using static IP addresses assigned to specific computers, manually configure the TCP/IP properties of each Windows NT computer with IP address and DNS information.

Windows NT allows only one TCP/IP stack configuration profile for LAN-to-Internet (or other network) connections. This means if you have more than one ISP connection available, you must change the settings in the TCP/IP Properties dialog box each time and then reboot the system. A shareware program called NetSwitcher (www.netswitcher.com) allows you to create multiple TCP/IP network connections and switch between them as needed. For example, if you're using a laptop and travel between different offices, each with its own LAN-based Internet access service, you can use NetSwitcher to create profiles for all your TCP/IP network connections and switch to the one you need when you're at a specific office. NetSwitcher also makes it easy to switch back and forth between a dedicated DSL connection and a dial-up connection when on the road.

PCI DSL modems are NICs in Windows NT

When you install a PCI DSL modem into your Windows NT computer, the device is installed as a network adapter. From your PC's perspective, the PCI DSL modem appears as a network interface card (NIC).

If your DSL service uses a PCI modem without PPP over DSL, you configure the TCP/IP settings for your computer by selecting the DSL modem driver in the Network properties dialog box.

Configuring for dynamic IP addressing

Using a DSL router or other Internet-connection-sharing solution that acts as a DHCP server enables dynamic IP addressing. This type of IP addressing is commonly used for consumer DSL offerings. You might get your IP address dynamically assigned by the ISP for your DSL connection. Or you might get only one IP address, which means you need to use a router or another Internet-sharing solution to share the DSL connection across multiple computers.

The DHCP server assigns an IP address from an available pool of IP addresses for use only during the current connection. The DHCP server functionality of the router allows IP addresses, subnet masks, and default gateway addresses to be assigned to computers on your LAN. You can use DHCP with registered IP addresses or private IP addresses along with the NAT (network address translation) feature incorporated in most DSL routers. Chapter 7 explains IP addressing and DHCP in more detail.

Each Windows NT computer on your LAN with an installed Microsoft TCP/IP stack is a DHCP client out of the box. If you've installed TCP/IP for the first time, you don't need to change any default settings to support DHCP. If you're unsure of your TCP/IP configuration, however, you can double-check to make sure the default TCP/IP settings are in place by performing the following steps:

1. **Right-click the Network Neighborhood icon and then choose Properties from the pop-up menu.**

   The Network properties dialog box appears.

2. **Click the Protocols tab.**

3. **Select the TCP/IP option and then click the Properties button.**

4. **Click the IP Address tab. If necessary, select the adapter card if you're using more than one card.**

5. **Select the Obtain an IP Address from a DHCP Server option.**

6. **Click OK twice to exit the Network applet.**

7. **Restart Windows NT.**

After you configure your Windows clients for DHCP, whenever you boot the Windows NT computer, it will get the IP address information it needs from the DHCP server.

Configuring for static IP addressing

If you plan on using registered IP addresses that you want to assign to specific computers — to run a Web server, for example — you need to configure the Windows NT computer with static IP address and DNS information. You can also use private (unregistered) IP addresses for your static IP addressing, but in most cases, you use private IP addresses with NAT and DHCP.

Before you configure your Windows NT computers for static IP addressing, you need to get the following information from your ISP:

- An IP address for each Windows NT computer
- A subnet mask IP address for your network
- The gateway IP address, which is the IP address assigned to your DSL router or your ISP's router if you're using a DSL modem
- A host and domain name for registered IP addresses
- DNS server IP addresses

To set up a Windows NT (Server or Workstation) for static IP addressing, do the following:

1. **In the Control Panel, double-click the Network icon.**

   The Network dialog box appears.

2. **Click the Protocols tab, select the TCP/IP Protocol option, and then click Properties.**

   The Microsoft TCP/IP Properties dialog box appears, as shown in Figure 13-3. If you have only one network adapter installed in your computer, Windows NT displays the adapter name in the Adapter list.

3. **If you have more than one adapter installed, use the drop-down list to select the adapter for which you want to configure TCP/IP properties.**

4. **Click the IP Address tab and then specify an IP address option.**

   The IP Address and Subnet Mask fields become active.

5. **Type the IP address and subnet mask IP address in their respective boxes.**

6. **In the Default gateway field, type the IP address of your DSL router or the ISP's router, if you're using a DSL modem (bridge).**

7. **Click the DNS tab.**

8. **In the Host Name box, type the host name of the client. In the Domain box, type the domain name.**

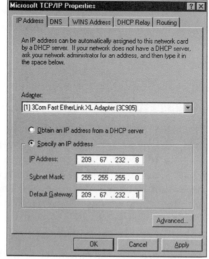

Figure 13-3:
The
Microsoft
TCP/IP
Properties
dialog box.

9. Type the IP address of the ISP's DNS server and then click the Add button. Repeat for each additional DNS server address supplied by your ISP.

10. Click OK twice to exit the Microsoft TCP/IP Properties and Network dialog boxes.

11. If prompted, insert your distribution disk or CD-ROM.

12. When prompted, click Yes to restart Windows NT.

Beyond basic TCP/IP configuration

Windows NT has more sophisticated TCP/IP capabilities than those in Windows 95 and 98. Windows NT's advanced IP addressing includes the following:

- Adding multiple IP addresses to a single NIC
- Choosing from multiple gateways
- Enabling a Microsoft virtual private network (VPN) solution
- Specifying TCP/IP security parameters

You can configure Windows NT's advanced IP addressing features by doing the following:

1. **Right-click the Network Neighborhood icon and then choose Properties.**

2. **Click the Protocols tab, select TCP/IP Protocol, and then click the Properties button.**

3. **Click the Advanced button.**

   The Advanced IP Addressing dialog box appears, as shown in Figure 13-4.

4. **Use the Advanced IP Addressing dialog box to do any of the following:**

   - In the IP Addresses area, use the Add, Edit, and Remove buttons to modify the IP addresses assigned to the selected network adapter. You can specify as many as 16 IP addresses per network adapter. You can use multiple IP addresses to make a single computer appear to the network as multiple virtual computers. This is useful if you want to run multiple independent Web servers on a single computer, with each Web server discretely addressable.

   - In the Gateways area, use the Add, Edit, and Remove buttons to modify the gateway's configuration. Gateways are searched in the order in which they're listed in the Gateways list (from top to bottom). If more than one gateway is listed, you can use the Up and Down buttons to modify the search order listing.

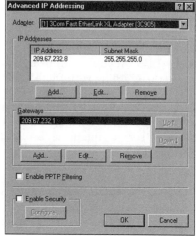

Figure 13-4:
The
Advanced
IP
Addressing
dialog box.

- Click to put a check mark in the Enable PPTP Filtering option, and you activate Microsoft's virtual private network (VPN) solution, called Point-to-Point Tunneling Protocol (PPTP). See Chapter 19 for more on VPNs.

- Click to put a check mark in the Enable Security option, and you can specify TCP/IP security parameters by clicking the Configure button to display the TCP/IP Security dialog box (see Figure 13-5). You can specify TCP Ports, UDP ports, and IP protocols on your computer that are available to the network. By default, all ports and protocols are available. You might, for example, set these parameters to allow other computers to access a Web server running on your computer while prohibiting them from accessing other TCP/IP services.

5. **After you've finished making changes in the Advanced IP Addressing dialog box, click OK to return to the Microsoft TCP/IP Properties dialog box.**

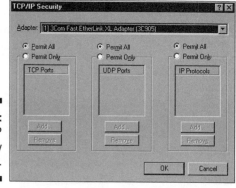

Figure 13-5:
The TCP/IP
Security
dialog box.

Working the DUN Way

If you're using a PCI or USB DSL modem that uses PPP over DSL, you need to use Windows NT Dial-Up Networking (DUN). If you're not using DSL CPE that uses PPP over DSL, you don't have to do anything with Dial-Up Networking.

Depending on the your DSL service, your Windows NT computer might be configured for DUN and TCP/IP as part of the service installation. Or the installation software provided with your service might have automatically configured the DUN and TCP/IP settings for you.

The Windows NT Remote Access Service (RAS) is the basis for any dial-up connection. RAS includes a Dial-Up Networking (DUN) facility similar to the one used by Windows 95/98, although it does have differences in implementation. The NT version of DUN is fused to RAS on the client side. Dial-Up Networking replaces the dial-out component of RAS, but during the installation and configuration of DUN, you'll see dialog boxes and references connected to RAS.

If you need to use Dial-Up Networking, you must first install RAS — if it's not installed already. After you install and configure RAS for DUN, you then create a connection profile. The DUN connection profile provides Windows NT with the information it needs to make the DSL connection. Some DSL modem vendors create the profile automatically as part of the installation, which uses information added to the software by the DSL Internet service provider.

The creation of a DUN connection profile mimics the creation of a DUN profile for an analog or ISDN modem — with one major difference. The entry for the telephone number is a Virtual Circuit (or Channel) Identifier (VCI) number. This 16-bit number is provided by the DSL ISP and identifies the channel your DSL service will use to make the connection.

Installing and configuring RAS for DUN

When you install a PCI DSL modem, the installation software typically installs Remote Access Service (RAS) if you haven't installed it already. You can check to see whether RAS is installed by checking the Services page in the Windows NT Network properties dialog box (see Figure 13-1). If RAS is installed, it appears as an entry in the Services page.

If RAS isn't installed, you must install it from your original Windows NT distribution media (CD-ROM) or from the hard drive (if you copied the original files there). To install and configure RAS, do the following:

1. **Open the My Computer folder on the desktop and then double-click the Dial-Up Networking icon.**

The Dial-Up Networking dialog box appears, asking whether you want to install the software.

2. **Click Install.**

The Windows NT Setup dialog box appears.

3. **Enter the location of your Windows NT original media.**

This is usually D:\i386, where D is the drive letter of your CD-ROM and i386 is the directory for Windows NT files for Intel platforms. If you copied the files to your hard drive, use the appropriate directory path instead of the path to your CD-ROM.

4. **Click Continue.**

The necessary files are transferred to your hard drive. Then the Add RAS Device dialog box automatically appears, and any ports that contain devices that can be used for DUN are in the RAS Capable Devices list.

5. **In the RAS Capable Devices list, select the DSL modem and then click OK.**

The Remote Access Setup dialog box appears.

6. **Select the Port/Device setting and then click the Configure button.**

The Configure Port Usage dialog box appears.

7. **Click to add a check mark to the Dial out only option and then click OK.**

8. **Click Close to exit the Network dialog box.**

9. **Restart Windows.**

Creating a Phonebook entry

Check your Dial-Up Networking window before you create a Dial-Up Networking profile. If you installed the PPP over DSL CPE, a DUN profile might have been created for you. If not, you can create a Dial-Up Networking profile using the Dial-Up Networking icon or the Internet Connection Wizard. A DUN profile includes all the specific IP address information you need to make a connection to the ISP.

You create a DUN profile as a Phonebook entry using the Dial-Up Networking icon:

1. **Double-click the My Computer icon on the desktop and then double-click the Dial Up Networking icon.**

The first time you open the Phonebook, a message box appears telling you that the phonebook is empty.

2. **Click OK.**

   The New Phonebook Entry Wizard appears.

3. **Type a name to identify your phonebook entry and then click Next.**

4. **Select the I am calling the Internet option and then click Next.**

5. **In the Phone Number field, enter a telephone number or the Virtual Circuit Identifier number, and then click Next.**

   Although DUN doesn't use the telephone number to make your DSL connection, you can't complete the connection profile unless you enter a number.

6. **Click Finish.**

   The Dial-Up Networking dialog box appears, as shown in Figure 13-6. The next time you double-click the Dial-Up Networking icon in the My Computer folder (after you create your first Phonebook entry), the Dial-Up Networking dialog box appears instead of the New Phonebook Entry Wizard.

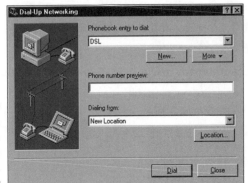

Figure 13-6:
The Dial-Up
Networking
dialog box.

Configuring your Phonebook entry for TCP/IP

After you create a Phonebook entry, you configure it for TCP/IP networking so that you can use it to connect to the Internet through the DSL modem and DSL connection. You use the Server properties page to configure TCP/IP for the Phonebook entry. This page includes a list of protocols for dial-up connections.

You configure the TCP/IP protocol for Internet connections, as follows:

1. **In the Dial-Up Networking dialog box, click the More button.**

   A menu appears.

2. **From the menu, choose Edit Entry and Modem Properties.**

   The Edit Phonebook Entry dialog box appears.

3. **Click the Server tab to display the server properties, as shown in Figure 13-7.**

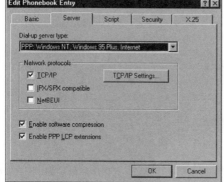

Figure 13-7:
The Server
tab in
the Edit
Phonebook
Entry
dialog box.

4. **In the list, select** *PPP: Windows NT, Windows 95 Plus, Internet.*

5. **In the Network protocols group, select TCP/IP.**

   Make sure the other two options in the group are not selected.

6. **If you're using DHCP, click OK to return to the Dial-Up Networking dialog box.**

7. **If you're using a static IP address, follow these instructions:**

   a. **Click the TCP/IP Settings button.**

      The PPP TCP/IP Settings dialog box appears, as shown in Figure 13-8.

   b. **Select the Specify an IP address option and then enter the IP address for the computer.**

   c. **Select the Specify name server addresses option and then enter the primary and secondary DNS server IP addresses.**

   d. **Click OK twice to return to the Dial-Up Networking dialog box.**

Keeping up appearances

Because DUN is designed primarily for dial-up sequences used by modems, it displays a number of message boxes to keep you apprised of the status of your dial-up connection. In most cases, you want to shut off these displays because the connection is instant and you don't want to be prompted for a response each time you make a connection.

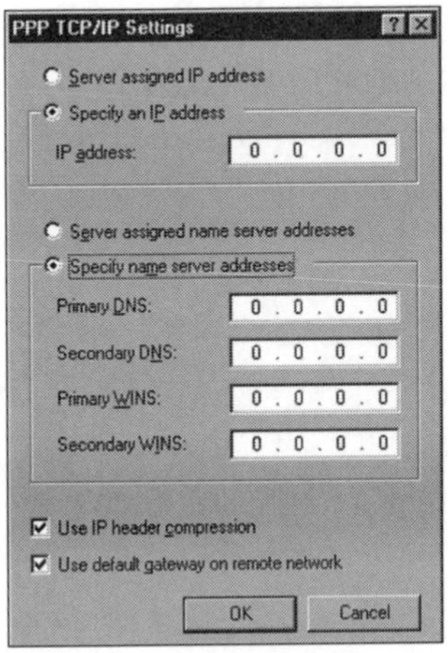

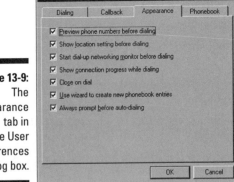

Figure 13-8:
The PPP
TCP/IP
Settings
dialog box.

In the Dial-Up Networking dialog box, click the More button and choose User preferences to display the User Preferences dialog box. For DSL dial-up connections, the only relevant item in the User Preferences dialog box is the Appearance tab, shown in Figure 13-9.

The Appearance tab allows you to set several options that control the appearance and function of DUN. Table 13-1 lists these settings and how you should configure them for working with a DSL connection.

Figure 13-9:
The
Appearance
tab in
the User
Preferences
dialog box.

Table 13-1	Settings in the Appearance Tab of the User Preferences Dialog Box	
Option	*Recommended Setting*	*What It Does*
Preview phone numbers before dialing	Unchecked	Displays and allows you to change the telephone number to be dialed before dialing occurs
Show location setting before dialing	Unchecked	Displays and allows you to change the location before dialing occurs
Start dial-up networking monitor before dialing	Unchecked	Activates DUN monitor automatically each time a DUN session is started
Show connection while dialing progress	Uncheck	Displays the progress of each DUN call step-by-step as the connection occurs and the session is established
Close on dial	Checked	Closes the Dial-Up Networking dialog box after dialing commences
Use Wizard to create phone-book entries	Checked/Unchecked	Uses the Dial-Up Networking Wizard to create new Phonebook entries; unchecking this setting lets you create new DUN entries using the standard dialog box instead of the Wizard
Always prompt before auto-dialing	Unchecked	Prompts you before auto-dialing a DUN Phonebook entry

Making the DUN connection

After you've created your Dial-Up Networking Phonebook entry, you're ready to make your connection. You can initiate a DUN connection in two ways. One way is to simply double-click the Web browser icon. The browser window opens along with the Connect To dialog box. Click the Connect button. The connection is made, and the home page appears in your Web browser. To disconnect, click the minimized Dial-up Connection dialog box on the taskbar and then click Disconnect.

Another way to make the Dial-Up Networking connection is shown in the following steps:

1. **In the My Computer folder, double-click the Dial-Up Networking icon.**

2. **In the Dial-Up Networking dialog box, click the Dial button.**

   The Dial-up Connection dialog box appears, as shown in Figure 13-10.

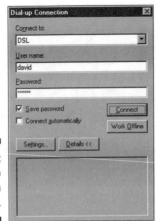

Figure 13-10:
The Dial-up Connection dialog box.

3. **Type your username and password for logging on to the Internet, if necessary.**

4. **Check the Save Password box to save the password so you don't have to enter it each time.**

   You can add or change your username and password using the Internet Properties dialog box.

5. **Click the Settings button.**

   The Dial-Up Settings dialog box appears.

6. **Click the Connect button to make your connection.**

You can monitor connections using the Dial-Up Networking Monitor, which is present on the status bar for the duration of the connection. To invoke the Monitor, double-click its icon. The Dial-Up Networking Monitor window appears. To end your connection, click Disconnect.

Handy Tools of the IP Trade

Windows NT includes two handy TCP/IP utilities — ipconfig and ping — that help you manage and monitor your IP connection. This section explains how to use these TCP/IP utilities.

You can find free or shareware Windows-based programs that provide a friendlier interface to the Windows NT TCP/IP utilities at popular software download sites such as www.download.com or www.zdnet.com/download.

What's your IP status?

The ipconfig program is available in Windows NT as a command-line program, although the Windows NT Resource Kit includes a graphics version similar to the one in Windows 95 and 98. This TCP/IP utility lets you view information about your TCP/IP protocol and network adapter settings. If you're using DHCP on your LAN to receive an IP address from a host, the ipconfig program is a handy way to know the IP address that has been assigned to your PC and how long it's been assigned.

To use the ipconfig program, choose Start ➪ Programs ➪ Command Prompt. The Command Prompt window appears. Typing **ipconfig** displays just the IP address, the subnet mask, and default gateway information, as shown in Figure 13-11. Typing **ipconfig/all** lists all IP address attributes. You can type **ipconfig/?** to display a list of all the commands you can use with ipconfig.

Figure 13-11: The results of entering the ipconfig command in the Command Prompt window.

```
Command Prompt                                              _ □ ✕
Microsoft(R) Windows NT(TM)
(C) Copyright 1985-1996 Microsoft Corp.

E:\>ipconfig

Windows NT IP Configuration

Ethernet adapter E190x1:

        IP Address. . . . . . . . : 192.168.168.200
        Subnet Mask . . . . . . . : 255.255.255.0
        Default Gateway . . . . . :

Ethernet adapter NdisWan6:

        IP Address. . . . . . . . : 0.0.0.0
        Subnet Mask . . . . . . . : 0.0.0.0
        Default Gateway . . . . . :

E:\>
```

Ping, ping, ping

The ping program is a handy TCP/IP diagnostic utility that works like TCP/IP sonar: You send a packet to a remote host and, if the host is functioning properly, it bounces the packet back to you. Ping prints the results of each packet transmission on your screen. By default, Ping sends four packets, but you can use it to transmit any number of packets or transmit continuously until you terminate the command.

You can access ping from the Command Prompt window (Start ➪ Programs ➪ Command Prompt). At the DOS prompt, enter the command in the following format:

```
ping IPaddress
```

where *IPaddress* is the IP address of the remote host you want to check. For example, if you type the following:

```
ping 199.232.255.113
```

and press Enter, you ping the host machine with the IP address of 199.232.255.113. The host machine can be on your local network or on the Internet. Likewise, you can ping the host name instead of the IP address for sites for which you don't know the IP address. For example, entering the following:

```
ping www.angell.com
```

returns the same ping information that you would get using the IP address. Figure 13-12 shows a sample ping result as it appears in the Command Prompt window.

```
Command Prompt                                              _ □ ×
Microsoft(R) Windows NT(TM)
(C) Copyright 1985-1996 Microsoft Corp.

D:\>ping 209.67.232.1

Pinging 209.67.232.1 with 32 bytes of data:

Reply from 209.67.232.1: bytes=32 time=50ms TTL=46
Reply from 209.67.232.1: bytes=32 time=50ms TTL=48
Reply from 209.67.232.1: bytes=32 time=50ms TTL=48
Reply from 209.67.232.1: bytes=32 time=50ms TTL=48

D:\>
```

Figure 13-12:
The
Command
Prompt
window with
ping results.

If the packets don't come back after you execute the ping command, either the host is not available or something is wrong with the connection.

The ping program is the single most useful program for troubleshooting the status of any TCP/IP connection. The ping command includes a number of parameters, which you can view by typing **ping** (in the Command Prompt window) and pressing Enter.

Chapter 14

Windows 2000 Meets DSL

*T*o use a computer running Windows 2000 (Professional or Server) to access the Internet through DSL, you need to configure the Microsoft TCP/IP This chapter explains the fundamentals of configuring Windows 2000 TCP/IP for connecting to the Internet through your DSL CPE.

If your Windows 2000 computer is configured as part of your DSL service, you can skip this chapter.

Windows 2000 and DSL Connections

As mentioned in Chapter 8, you can use different types of DSL CPE as part of your Internet service You can use different types of DSL CPE as part of your Internet service. (For more on DSL CPE, check out Chapter 8.) How the different types of DSL CPE interface to Windows 2000 varies.

Depending on the DSL Internet service package you choose, the DSL provider might configure your Windows 2000 computer for TCP/IP as part of the service installation or provide installation software that automatically configures your TCP/IP settings for you. If your Windows 2000 computer is configured as part of your DSL service, you can skip this chapter.

Here is how different DSL CPE interface with Windows 2000 and where you configure their TCP/IP settings:

- ✔ If you use **an Ethernet DSL modem or router,** Windows 2000 uses the Ethernet interface through a NIC (network interface card). Using an Ethernet solution means that you configure the NIC card for TCP/IP.

- ✔ If you use **a DSL PCI or USB modem that uses PPP over DSL,** Windows 2000 uses Dial-up Connections to make the Internet connection through DSL.

- ✔ If you use **a DSL PCI or USB modem that doesn't use PPP over DSL**, Windows 2000 uses the Ethernet interface. You configure TCP/IP for the DSL CPE device driver as you would configure a NIC.

Setting Up TCP/IP in Windows 2000

The first step in configuring Windows 2000 for TCP/IP is to make sure the Microsoft TCP/IP stack is installed. If you've been connecting to the Internet already or installed a NIC, the Microsoft TCP/IP stack is already installed on your computer.

You can check to see whether the TCP/IP stack is installed by doing the following:

1. **Right-click the My Network Places icon on the desktop and choose Properties from the pop-up menu**.

2. **Right-click the Local Area Connection icon and select Properties from the pop-up menu**.

3. **In the Local Area Connection dialog box, check for the Internet Protocol (TCP/IP) entry.**

   If you see the entry, TCP/IP is installed.

If TCP/IP isn't installed, you need to install the Microsoft TCP/IP stack.

When TCP/IP is installed on your computer, Windows 2000 binds the TCP/IP protocol to your NIC and to the Dial-up Connection used for PPP over DSL connections. *Binding* is a process that defines the relationship between a network adapter card and network protocols used with that network card. When you bind the TCP/IP protocol to a network adapter, you allow TCP/IP traffic to pass through your Ethernet network to the DSL modem or router. Likewise, binding TCP/IP to the Dial-up Connection profile for your DSL over PPP Internet connection enables you to use TCP/IP for DSL connections using Dial-up Connections.

To install the Microsoft TCP/IP stack, do the following:

1. **Right-click the My Network Places icon on the desktop and choose Properties from the pop-up menu.**

   The Network and Dial-up Connections window appears

2. **Right-click the Local Area Connection icon and select Properties from the pop-up menu**.

   The Local Area Connection Properties dialog box appears.

3. **Click Install.**

   The Select Network Component Type dialog box appears.

4. **Double-click Protocol.**

5. **Select Internet Protocol (TCP/IP) and then click OK.**

6. **Click Close to complete the installation.**

Congratulations, you just installed the Microsoft TCP/IP software on your computer.

Configuring Your NICs for TCP/IP

When you bind the TCP/IP protocol to a network adapter, you allow TCP/IP traffic to pass through your Ethernet network to the DSL bridge or router. You need to configure the Microsoft TCP/IP stack bound to the NIC for the IP address configuration you're using on your LAN.

You can configure your TCP/IP stack in two ways:

- ✔ If you're using a DSL router as a DHCP (Dynamic Host Configuration Protocol) server to dynamically assign IP addresses, configure your Windows 2000 computers as DHCP clients and don't add any specific IP address information to the TCP/IP properties.

- ✔ If you're using static IP addresses assigned to specific computers, manually configure the TCP/IP properties of each Windows 2000 computer with IP address and DNS information.

 Windows 2000 allows only one TCP/IP stack configuration profile for LAN-to-Internet (or other network) connections. This means if you have more than one ISP connection available, you must change the settings in the TCP/IP Properties dialog box each time you switch to a different TCP/IP network and then reboot the system. A shareware program called NetSwitcher (www.netswitcher.com) allows you to create multiple TCP/IP network connections and switch between them as needed. For example, if you're using a laptop and travel between different offices, each with its own LAN-based

Internet access service, you can use NetSwitcher to create profiles for all your TCP/IP network connections and switch to the one you need when you're at a specific office. NetSwitcher also makes it easy to switch between a dedicated DSL connection and a dial-up connection when you're on the road.

Accessing Windows 2000 TCP/IP properties

In Windows 2000, you use the Internet Protocol (TCP/IP) Properties dialog box to configure your TCP/IP properties for NICs as well as configure PCI and USB modems that don't use PPP over DSL. This is the dialog box you use to configure the TCP/IP properties for DSL Internet service.

To access the Internet Protocol (TCP/IP) Properties dialog box, do the following:

1. **Right-click the My Network Places icon on the desktop and choose Properties from the pop-up menu**.

   You can also choose Start ⇨ Settings ⇨ Network and Dial-Up Connections. The Network and Dial-up Connections window appears, as shown in Figure 14-1.

Figure 14-1:
The Network and Dial-up Connections window is the gateway to configuring all your LAN and Internet connections in Windows 2000.

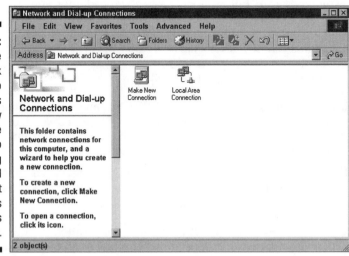

2. **Right-click the Local Area Connection icon and then choose Properties from the pop-up menu**.

   The Local Area Connection Properties dialog box appears, as shown in Figure 14-2.

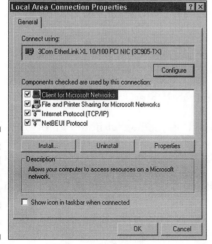

Figure 14-2:
The Local
Area
Connection
Properties
dialog box.

3. **In the list in the middle of the dialog box, double-click the Internet Protocol (TCP/IP) entry.**

   You can also select Internet Protocol (TCP/IP) and then click the Properties button The Internet Protocol (TCP/IP) Properties dialog box appears, as shown in Figure 14-3.

In the Internet Protocol (TCP/IP) Properties dialog box, you configure Windows 2000 for static or dynamic IP addressing. See "Configuring for dynamic IP addressing" or "Configuring for static IP addressing" later in this chapter for specific instructions.

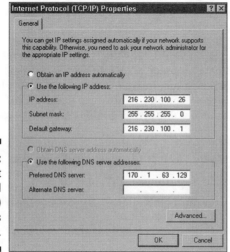

Figure 14-3:
The Internet
Protocol
(TCP/IP)
Properties
dialog box.

USB and PCI DSL modems are NICs in Windows 2000

When you install a USB or PCI DSL modem into your Windows 2000 machine, it's installed as a network adapter. From your PC's perspective, the USB or PCI DSL modem appears as a network interface card (NIC).

If your DSL service uses a PCI or USB modem without PPP over DSL, you configure the TCP/IP settings for your computer using the Internet Protocol (TCP/IP) Properties dialog box in the same way you do for a NIC connection. The only difference is that instead of selecting the NIC adapter in the Local Area Connection Properties dialog box, you select the DSL modem driver that's bound to TCP/IP.

Configuring for dynamic IP addressing

Using a DSL router or other Internet-connection-sharing solution that acts as a DHCP server enables dynamic IP addressing. This type of IP addressing is commonly used for consumer DSL offerings. You might get your IP address dynamically assigned by the ISP for your DSL connection. Or you might get only one IP address, which means you need to use a router or another Internet-sharing solution to share the DSL connection across multiple computers.

The DHCP server assigns an IP address from an available pool of IP addresses for use only during the current connection. The DHCP server functionality of the router allows IP addresses, subnet masks, and default gateway addresses to be assigned to computers on your LAN. You can use DHCP with registered IP addresses or private IP addresses along with the NAT (network address translation) feature incorporated in most DSL routers. Chapter 7 explains IP addressing and DHCP in more detail.

Each Windows 2000 computer on your LAN with TCP/IP installed is a DHCP client out of the box. If you haven't made any changes to your default TCP/IP settings, you don't need to change any default settings to support DHCP. If you're unsure of your TCP/IP configuration, however, you can double-check to make sure your Windows 2000 computer is configured for being a DHCP client:

1. **Open the Internet Protocol (TCP/IP) Properties dialog box.**

2. **Select Obtain an IP Address automatically, if necessary.**

3. **Select Obtain DNS server address automatically, if necessary.**

4. **Click OK twice to exit.**

Whenever you boot the Windows NT computer (after you configure your Windows clients for DHCP), it will get the IP address information it needs from the DHCP server.

Configuring for static IP addressing

If you plan on using registered IP addresses that you want to assign to specific computers — to run a Web server, for example — you need to configure the Windows NT computer with static IP address and DNS information. You can also use private (unregistered) IP addresses for your static IP addressing, but in most cases, you use private IP addresses with NAT and DHCP.

Before you configure your Windows 2000 computers for static IP addressing, you need to get the following information from your ISP:

- An IP address for each Windows NT computer
- A subnet mask IP address for your network
- The gateway IP address
- DNS server IP addresses

To set up a Windows 2000 for static IP addressing, do the following:

1. **Open the Internet Protocol (TCP/IP) Properties dialog box.**
2. **Select Use the following IP address.**

   All the fields below become active.
3. **In the IP address field, enter the IP address you're assigning to the computer. In the Subnet mask field, enter the subnet mask IP address.**
4. **In the Default gateway field, enter the IP address of the router you're using to connect to the Internet.**

   If you're using a router at your end of the DSL connection, the gateway IP address is the one assigned to your router or PC acting as the gateway. If you're using a DSL modem, which is a bridge, the gateway IP address is your ISP's router on their end of the IP connection.
5. **In the Preferred DNS server field, enter the IP addresses of the primary DNS server.**

   If you're ISP provided you with another DNS server IP address, enter it in the Alternate DNS server.
6. **Click OK twice to complete the installation.**

Advanced TCP/IP settings

You can have multiple NICs installed in Windows 2000, each with its own TCP/IP configuration settings. Clicking the Advanced button in the Internet Protocol (TCP/IP) Properties dialog box, displays the Advanced TCP/IP Settings dialog box (Figure 14-4)

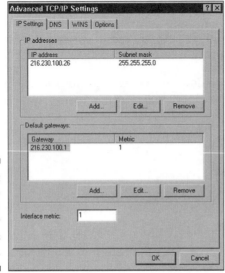

Figure 14-4:
The
Advanced
TCP/IP
Settings
dialog box.

On the IP Settings page, you can enter additional IP addresses for any installed network adapter. Assigning multiple IP addresses to a network adapter allows the computer to run multiple servers that share the same NIC interface. This means you can run multiple Web servers, for example, on a single computer with each server having its own IP address and host name. You can also specify the addresses of additional gateways for each adapter in the IP Settings page. Unlike the additional IP addresses, which all remain active simultaneously, additional gateways are only used in the order they're listed, when the primary default gateway is unreachable.

The DNS page allows you specify DNS server IP addresses in the order of use. These settings are useful, if you're running a DNS server on your LAN to support an Intranet.

The Options page allows includes settings for configuring IP Security and TCP/IP filtering. You can specify to use IPSec to your Internet Protocol connections to provide security and privacy for data being sent over your Internet Protocol connection. TCP/IP filtering is enabled for your network adapter to provide security.

Setting Up Dial-up Connections

If you're using a PCI or USB DSL modem that uses PPP over DSL, you need to use Windows 2000 Dial-up Connections. If you're not using DSL CPE that uses PPP over DSL, you don't have to do anything with Dial-up Connections.

Depending on the your DSL service, your Windows 2000 computer might be configured for Dial-up Connections and TCP/IP as part of the service installation or the installation software provided with your service automatically configures the Dial-up Connections and TCP/IP settings for you.

A Dial-up Connection profile provides Windows 2000 with the information it needs to make the DSL connection to the Internet. The creation of a Dial-up Connection profile mimics the creation of a Dial-up Connection profile for an analog or ISDN modem — with one major difference. The entry for the telephone number is a virtual circuit identifier (VCI) number. This 16-bit number is provided by the DSL ISP and identifies the channel your DSL service will use to make the connection.

Making the dial-up connection

You use the Network Connection Wizard to create a new Dial-up Connection profile for your PPP over DSL connection.

1. **In the Network and Dial-up Connections window, double-click the Make New Connection icon.**

   The Network Connection Wizard appears.

2. **Click Next to display the Network Connection Type page, as shown in Figure 14-5.**

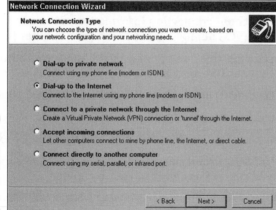

Figure 14-5: The Network Connection Type page.

3. Click the Dial-up to the Internet option and then click Next.

The Internet Connection Wizard appears, as shown in Figure 14-6.

Figure 14-6:
The Internet
Connection
Wizard.

4. Select the third option — I want to set up my Internet connection manually, or I want to connect through a local area network (LAN) — and then click Next.

The setting up your Internet connection page appears.

5. Select the *I connect through a phone line and a modem* option and then click Next.

6. If you have more than one modem installed, select your DSL modem in the Choose modem page and then click Next.

7. If the DSL modem is the only modem installed, the Internet account connection information appears.

8. In the Internet account connection information, enter your Virtual Circuit Identifier (VCI) number in the Telephone number field.

Your DSL ISP provides this number or it might be entered automatically when you select your DSL modem driver in the previous page.

9. If you're using static IP addresses, click on the Advanced button.

The Advanced Connection Properties dialog box appears.

10. Click the Addresses tab, do the following, and then click Next:

- In the IP address group, select Always use the following and then enter the IP address assigned to your computer.

- In the DNS server address group, select Always use the following and then enter the IP addresses for the DNS servers.

11. **In the Internet account login information page, enter the username and password for your Internet access account and then click Next.**

12. **Enter the name of your connection and then click Next.**

13. **Do one of the following:**

     - If you're setting up your email for the first time, click Yes, click the Next button, and then follow the instructions for the next few pages to configure your email.

     - If you already have an email account, select Yes, click Use an Existing Internet Mail, and then click the Next button twice.

     - If you want to bypass configuring email altogether and do it later through your email program, click No and then click the Next button.

14. **If you want to try out your connection immediately, make sure there's a check mark before *To connect to the Internet immediately, select this box and then click Finish*. Then click Finish.**

     Your default Web browser starts to test the connection, and the Connect dialog box appears.

15. **Click Dial to make your connection.**

After you create your Dial-up Connection profile for your DSL connection, create a desktop shortcut for the connection. Right-click on the connection icon and choose Create Shortcut from the pop-up menu.

Bypassing the Connect dialog box

Every time you use a Dial-up Connection profile, the Connect dialog box appears, prompting you for confirmation of your connection before making the call. You can get rid of this Connect dialog box and automatically make the dial-up connection by doing the following:

1. **In the Network and Dial-up Connections window, right-click the Dial-up Connection profile.**

   The properties window for the Dial-up Connection profile appears. If the name of your Dial-up Connection profile is DSL Internet, for example, the name of the dialog box is DSL Internet Properties.

2. **Click the Options tab.**

3. **Uncheck the Prompt for name and password, certificate, etc. option.**

4. **Click OK.**

The Properties dialog box includes all the settings to customize your Dial-up Connection profile. In the Networking tab, you can configure your TCP/IP settings and enable Internet-connection sharing.

Handy TCP/IP Utilities

Windows 2000 has a collection of TCP/IP utilities that include ipconfig and ping. These tools help you to manage and monitor your IP connection. These tools are the same ones in Windows NT. See Chapter 13 for more information on these utilities.

Chapter 15

The Mac Meets DSL

In This Chapter

▶ Connecting DSL Internet service to the Mac

▶ Discovering how to configure TCP/IP in the Mac OS

▶ Using the Internet Setup Assistant

▶ Configuring Internet applications from the Internet control panel

Apple has fully integrated networking into its reborn Mac product line to make connecting through an Ethernet modem or router easy. After your DSL CPE is in place, you simply configure your network adapter in the Mac OS for TCP/IP.

This chapter explains the fundamentals of configuring the Mac OS for connecting to the Internet.

Mac and DSL Connections

iMacs, iBooks, PowerBooks, and Power Mac G4 all have built-in 10/100 Ethernet adapters. Most also support Apple's AirPort 11-Mbps wireless networking. This chapter focuses on Mac OS 8.5, 8.6, and OS 9.0 in explaining how to configure Mac TCP/IP for Internet connections.

You can configure TCP/IP in the Mac using two methods. You can either manually enter your Internet connection information (IP addresses, DNS servers, and so on) into the TCP/IP control panel or use the Internet Setup Assistant.

Configuring TCP/IP Manually

Mac OS includes the TCP/IP control panel for manually setting up your Internet connection. It's a straightforward way to configure TCP/IP settings for Internet access through your network adapter. You can configure Mac OS TCP/IP to support either static IP addressing or dynamic IP addressing:

✔ If you're using static IP addresses, you configure the TCP/IP properties of the Mac with IP address and DNS information supplied by your ISP.

✔ If you're using a router as a DHCP (Dynamic Host Configuration Protocol) server to dynamically assign IP addresses, you configure your Mac as a DHCP client.

Configuring for static IP addresses

Before you configure your Macintosh clients for static IP addressing, you need to get the following information from your ISP:

✔ An IP address for each Macintosh

✔ A subnet mask IP address for your network

✔ The default gateway IP address

✔ DNS server primary and secondary IP addresses

✔ A domain name

Here's how to configure your Mac for static IP addressing:

1. **Choose Apple menu ⇨ Control Panels ⇨ TCP/IP.**

   The TCP/IP control panel appears, as shown in Figure 15-1, with the name of the active configuration in the title bar.

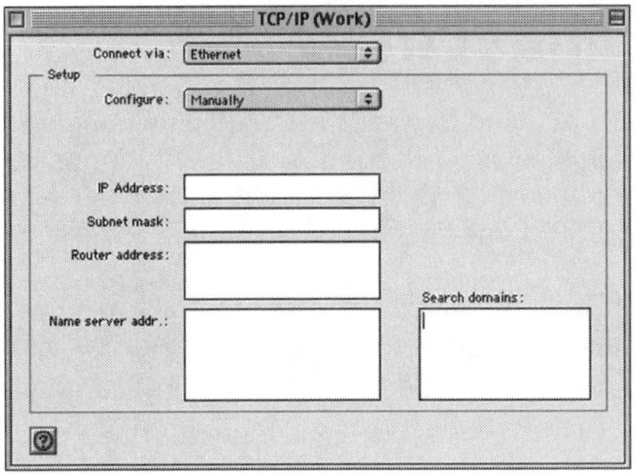

Figure 15-1:
The TCP/IP
control
panel.

2. **To select a different configuration, choose File ⇨ Configurations.**

3. **In the Configurations window, select the setting you want to use and then click Make Active.**

   If you plan to use this setting exclusively, choose the Default setting.

4. **In the Connect via list, choose Ethernet.**

5. **In the Configure list, choose Manually**.

6. **Enter the following information:**

   - In the IP Address box, enter the IP address for your Mac.

   - In the Subnet mask box, enter the subnet IP address for your network.

   - In the Router address box, enter the gateway IP address.

   - In the Name server addr box, enter the DNS server IP addresses.

   - In the Search domains box, enter the domain name you're using with your DSL Internet service.

7. **Close the TCP/IP control panel window.**

   A dialog box appears prompting you to save your configuration settings.

8. **Click Save to save your changes.**

Configuring for dynamic IP addresses

Here's how to configure your Macintosh for dynamic IP addressing:

1. **Choose Apple menu ⇨ Control Panels ⇨ TCP/IP.**

   The TCP/IP control panel appears with the name of the active configuration appears in the title bar.

2. **To select a different configuration, choose File ⇨ Configurations.**

3. **In the Configurations window, select the setting you want to use and then click Make Active.**

   If you plan to use this setting exclusively, choose the Default setting.

4. **In the Connect via list, choose Ethernet.**

5. **In the Configure list, choose DHCP.**

6. **Close the TCP/IP control panel window.**

   A dialog box appears prompting you to save your configuration settings.

7. **Click Save to save your changes.**

Configuring TCP/IP with Internet Setup Assistant

The Internet Setup Assistant is an application that steps you through the process of configuring Mac TCP/IP settings for your network adapter. While configuring your Mac for a DSL Internet connection, the Assistant lets you configure your email, network news, and other Internet connection settings in addition to configuring your IP address and DNS service information.

Here's how to use the Internet Setup Assistant:

1. **In the Assistants folder, double-click the Internet Setup Assistant icon.**

   The Internet Setup Assistant's main window appears.

2. **Click the Yes button.**

   The second window of the Internet Setup Assistant appears.

3. **Click Yes.**

   The Introduction window appears.

4. **Read the text and then click the right arrow at the bottom of the window.**

   The Configuration name and connection type window appears.

5. **In the edit box, enter a name for the connection, select Network, and then click the right arrow.**

   The IP Address window appears.

6. **If you've been assigned an IP address, select Yes. If you're using a dynamic IP address, select No.**

7. **If you selected Yes in Step 6, another IP Address window appears. Enter the IP address in the edit box and then click the right arrow.**

8. **If you selected Yes in Step 6, the Subnet Mask and Router Address window appears. Enter the subnet mask IP address and router IP address, and then click the right arrow.**

   The router IP address is either the IP address of the DSL router you're using or the IP address of the ISP's router on your DSL modem connection.

9. **In the Domain Name Servers window, enter the DNS address or addresses in the edit box.**

   If you need to enter more than one address, make sure you press Return between each one.

10. **In the bottom edit box, enter the domain name and then click the right arrow.**

    The E-mail Address and Password window appears.

11. **In the top box, enter your email address; in the bottom box, enter your email password. Then click the right arrow.**

    The E-mail Account and Host Computer window appears.

12. **Enter the ISP's POP account and SMTP host information and then click the right arrow.**

    The Newsgroup Host Computer window appears.

13. **Enter the NNTP host and then click the right arrow.**

    The Proxies window appears.

14. **If you're using a proxy server for the Internet connection, select Yes; otherwise, click No. For either case, click the right arrow.**

15. **If you selected the Yes option in Step 14, click to place a check mark in the boxes for the types of proxies you use, enter the corresponding proxy information, and then click the right arrow.**

    The Conclusion window appears.

16. **Click Go Ahead.**

    The Internet Setup Assistant window indicates the progress of your configuration. When it's finished, it quits.

Taking Control through the Internet Control Panel

The Internet control panel enables you to set a variety of application configuration options, including personal email, Web, and news information. Many of these options are configured if you use the Internet Setup Assistant.

Microsoft Outlook and Microsoft Internet Explorer are the email program and Web browser, respectively, bundled with Mac OS. You can use the pop-up menus in the E-mail, Web, and News tabs to locate and choose other applications you've installed on your computer for Internet access. You can also use commands in the Internet control panel's File menu to create, duplicate, and rename a set of options.

To open the Internet control panel and configure options, do the following:

1. **Choose Apple menu ⇨ Control Panels ⇨ Internet.**

    The Internet control panel appears, as shown in Figure 15-2.

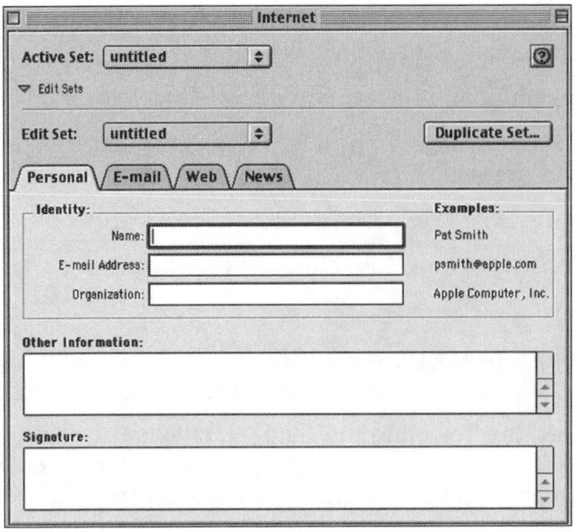

Figure 15-2:
The Internet control panel.

2. **To make a different set of options active, choose the set's name from the Active Set pop-up menu.**

3. **To edit a set of options, choose the set's name from the Edit Set pop-up menu.**

4. **Make changes to any of the options in the following four tabs:**

    • **Personal** includes your name, email address, and organization. You can also enter additional information and an email signature that can appear at the bottom of every email message you send.

    • **E-mail** includes your email configuration information as well as notification options for incoming email and your default email application.

    • **Web** includes the URLs for your default home and search pages as well as the folder for downloaded files, colors to use for Web page links and backgrounds, and your default Web browser application.

    • **News** includes your Internet news configuration information and your default newsreader application.

5. **Close the Internet control panel.**

    A confirmation dialog box appears.

6. **Click Save.**

You're ready to start using your DSL-powered Internet connection.

Part IV
Got DSL, Now What?

"THE IMAGE IS GETTING CLEARER NOW... I CAN ALMOST SEE IT... YES! THERE IT IS — THE GLITCH IS IN A FAULTY CABLE AT YOUR OFFICE IN DENVER."

In this part . . .

Time to go beyond the mechanics of DSL and get into the good stuff. What can you do with a DSL Internet connection? DSL is broadband but this means more than just faster Internet access. Broadband affects your home or business from the minute your service is turned on — and the uses and benefits of DSL service expand in ever-growing circles.

You look at how broadband integrates the Internet into your daily life at home. Next, you see how the DSL-powered business works smarter by tapping into all the Internet has to offer.

You discover how to include video conferencing and voice to your DSL connection to add show-and-tell to your Internet communications. You also find out how a business can use DSL with virtual private networking (VPN) to put telecommuters in the fast lane.

Chapter 16

The Broadband-Powered Home

*I*t's time to go beyond the mechanics of DSL and get to the fun stuff. Welcome to the broadband lifestyle!

Broadband is more than just a faster connection to the Internet — it unleashes the Internet's full potential. Your DSL-wired computer transforms into an essential household appliance for the whole family. This chapter guides you through the ways broadband can change your life.

Going Broadband

The broadband-powered household can make you a full participant in all the Internet has to offer. Over time, the Internet becomes an extension of your daily life and an essential source for all kinds of activities.

You can do all kinds of cool things on the Internet with DSL that you simply can't do with a clunky dial-up Internet connection, such as

✔ **Surf the Web at exhilarating speed.** The days of waiting for Web pages to appear on your screen are over. Fast Web surfing means you can do more in less time.

✔ **Play online.** High-speed DSL opens up all the multimedia entertainment richness the Internet has to offer. You can view real-time video broadcasts directly from the Internet or quickly download video files and play them on your computer. You can also listen to music files or broadcasts and make your own music CDs. The power of a high-speed connection shines when playing graphic-intensive online games.

✔ **Download software at lightening speed.** Get the latest updates and programs fast. No more long coffee breaks while you wait for files to download — a 5MB file takes 90 seconds to download! Whether you're buying a software update or downloading a free program from popular software sites such as www.zdnet.com or www.download.com, this is the way you get your software. You can even rent software from Application Service Providers (ASPs). ASP Web sites act as online software vending machines, where you rent software or online services, such as multiuser games and expensive specialized business applications.

✔ **Send and receive email with large file attachments.** Now you can quickly send big files — video clips, photos, and programs — as email attachments.

✔ **Get email messages instantly.** Always-on Internet access means you can get and send you email instantly. You hear a beep; you've got mail.

✔ **Do video conferencing.** Communicate face-to-face with family and friends over the Internet using inexpensive video conferencing kits. Although it's not the quality of the Jetson's, it's a fun way to keep in touch.

✔ **Transacting household business.** Shopping takes on a new dimension with a high-speed connection as you breeze through your favorite e-commerce Web sites. Your DSL connection also gives you full-powered access to online investing, bill paying, banking, and more.

✔ **Telecommuting.** DSL combined with VPN (virtual private networking) brings all the capabilities of your office network to any computer in your home.

Broadband-powered computing

The power of broadband makes you want to rearrange where you do your computing at home. Computers end up in new places, such as the living room, the kitchen, the bedroom, or even the deck. If you don't have a home network, you'll want one. Luckily, you can choose from a variety of home networking technologies. For the ultimate in networking, you might set up a wireless network to deliver cordless broadband Internet to every room in your home.

The effect of broadband carries over to how you outfit your computers. To start, you'll probably opt for maximum processing power to handle the intensive data processing of broadband-enabled applications such as video. A good sound system is essential for playing music, listening to Internet broadcasts, and creating your own music CDs. And perhaps you want an unobtrusive flat-panel display to substantially reduce the computer's footprint.

Living the broadband lifestyle

Your high-speed, always-on DSL connection integrates the Internet into your daily life. The broadband lifestyle is about convenience and freeing up time. Broadband Internet delivers cool life enhancements, such as the following:

- **Lets your fingers do the traveling.** High speed and instant access means you're using broadband for a growing number of tasks, such as shopping, banking, investing, and learning. You'll find yourself making far more frequent Internet trips.

- **Opens up new visual ways to communicate.** The power of your DSL connection can enhance the ways you keep in touch with friends, family, and colleagues beyond just email. Using a video conferencing kit, you can drop in for face-to-face conversations. Even your home videos can be instantly shared by sending them as self-running email attachments. Post the latest family pictures from your digital camera to share with family and friends through your own personal Web site.

- **Makes you a more productive telecommuter.** Broadband puts you in the fast lane for remote access to your company network. Your DSL connection allows you to work at home using a high-speed, always-on service like the one you're already using at the office — or maybe even faster than the one at your office!

Working smarter from home

The DSL-powered telecommuter works faster and smarter from home. Using broadband, you can work more productively at home with the bonus of capturing time usually lost to commuting on congested highways.

Your DSL connection combined with VPN puts you in the diamond lane of telecommuting. VPN technology allows private data to pass securely over the public Internet. The same connection you use for Internet access becomes your high-speed link to the office.

Through VPN, you have access to all the essential services at your office, including email, databases, files, and intranet resources. The high speed of DSL creates the same feel of your office network from the comfort of your home office. In addition, DSL helps you stay in the loop at the office through the use of sophisticated tools such as Web-based presentations and video conferencing. For more information on virtual private networking, check out Chapter 19.

A day in the life of a broadband household

Broadband makes the Internet a seamless part of everyday life. In this section, you take a walk through a day in the life of a typical broadband-powered household to show how a DSL connection can affect your daily life.

Beginning your day. Pulling your notebook from the nightstand, you power it up. The notebook is connected to a wireless network that connects to the Internet through DSL. You catch up on the news by checking out CNN and the New York Times online. Next, you check your personal email and then visit the family Web site to see digital pictures of your new nephew.

Getting to work. After taking that grueling commute down the hall to your home office, you sit down to telecommute. Double-clicking the VPN icon on your computer screen instantly secures your link over the Internet to your office network. Your home computer becomes part of the company network with full access to all essential office tools (email, calendar, files, and databases).

Restocking the fridge. At lunchtime, you head to the kitchen to check out your lunch options. The refrigerator is empty and the only thing around is a package of Ramen noodles. It's time to buy some groceries. At the kitchen table, you use your notebook, which is connected to DSL through the wireless network, to go to the Webvan site (Figure 16-1). As you eat your noodles, you order groceries using an electronic shopping list, check out, and then choose a delivery time.

Back at work. It's a bit later in the day, time for you to give a presentation to a group at the office using a Web-based presentation site. For the rest of the afternoon, you conduct research over the Web while listening to your favorite Internet radio broadcasts.

Kids do homework and playing music. The kids come home from school and connect to the Internet to get answers for a school project. Next, it's a visit to their favorite online game site. Before leaving, they do a quick download of music files into their portable MP3 music player.

Cooking with broadband. As your Webvan groceries arrive, you're searching for a recipe to do something new and exciting for dinner at the kitchen counter. You unpack your groceries and print a few recipes from Bon Appetit, Cooking Light, and iChef. Then you stock up on wine after getting a little education at the wine.com Web site.

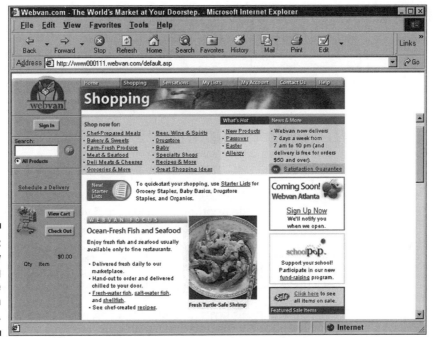

Figure 16-1:
Grocery
shopping
from home
with
Webvan.

After-hours shopping. It's called e-commerce, and it's convenience for you. You shop for a present for that new nephew and have it shipped from eToys. You've also had your eye on a Palm Pilot, so you decide to buy. You check PC Magazine for reviews of the different models; then it's on to comparison shopping, followed by a purchase at Buy.com for next-day delivery. A birthday is coming up for your sister, and you need a gift idea. Your sister likes clothes from Talbots, so you buy her a gift certificate at their site. They gift wrap it and ship it to her with your personal birthday wishes.

Dropping in on friends and family. You make a video conference call to your sister in Albuquerque and then to a friend in Boston. You can drop in for a quick video conference call or linger for as long as you want because toll charges aren't an issue. Your computer rings and its an incoming video call from a buddy in California.

Ending your day. You end your day with a quick visit to your family Web site, to post those pictures of your last vacation to St. Martin.

Broadband Power Tools

Cool Internet experiences await you with broadband: multimedia enhanced information, entertainment, shopping, up-to-the-minute news, live music, multiplayer games, virtual reality tours, and live sporting events, to name a few.

Multimedia moments

Before you embark on your broadband journey, you should take stock of your multimedia players. *Media players* are software programs that plug in to your browser (Microsoft Internet Explorer or Netscape Communicator) to let you view video, listen to music or other audio files, and view animated graphics. All these tools are free for the downloading.

Different media players support different file formats, so you might end up using more than one player as you use multimedia resources at different Web sites. Check these sites for the latest versions:

- **Real Networks' RealPlayer,** at `www.real.com`, is the most widely used multimedia player on the Internet for playing video and audio content. The latest version supports MP3, which is the universal file format for digital music on the Internet. (See the "Making music" section for more information on MP3.)

- **Microsoft Media Player,** at `www.microsoft.com/windows/mediaplayer`, is the second most popular multimedia player. Media Player is included with Windows, but you might need to upgrade it. The latest version of Windows Media Player plays MP3 music files Internet.

- **Apple's QuickTime Player,** at `www.apple.com/quicktime`, gives you a friendly interface with controls you'd expect to find on a television. You also get enhanced controls for storing and playing movies.

- **Macromedia's Shockwave Player,** at `www.macromedia.com/shockwave`, is used for viewing animated graphics as well as interactive Web content such as games, business presentations, and entertainment from your Web browser.

Making music

MP3, which stands for MPEG Layer 3, is a file format for storing near CD-quality music. MP3 is the most popular music file format on the Internet because it compresses high-quality music into relatively small file sizes.

Most media players support playing MP3 files, but the following MP3 players allow you to do a lot more with your MP3 music files, including creating customized music CDs (if you own a CD writer):

✔ **Winamp** is a power-packed MP3 player that gives you everything you need to play music, download music files, organize the files in sophisticated databases, create music CDs from downloaded files, mix music, and a lot more. Check out Figure 16-2. Winamp's Skins lets you change the look and feel of Winamp to your liking. It even includes a 10-band graphic equalizer to customize how your music sounds. Best of all, Winamp runs in a very small window that gives you quick access to all the controls. You can download the full version of Winamp player from www.winamp.com.

✔ **MusicMatch Jukebox** also lets you record your favorite tracks from your CD library into MP3, add new MP3 music from the Internet to your digital music collection, and organize your MP3 recordings into a sophisticated music database on your PC. If you have a CD writer, you can use MusicMatch Jukebox to create your own music CDs from downloaded music. You can download a stripped-down version of MusicMatch Jukebox from www.musicmatch.com. To get all the features, however, you need to purchase a copy for around $30.

Figure 16-2:
The
Winamp
music
player.

Going face to face

DSL-powered households can use video conferencing to drop in on other broadband-powered friends and family by simply calling them from your computer. Both your video and voice move over your flat-rate DSL connection. Although you don't get the quality shown on the Jetsons, video conferencing over a DSL connection is impressive.

A video conferencing kit includes a small digital camera with a built-in microphone that typically mounts on the top of your monitor, although you can place it anywhere. Most video conference kits are USB for easy installation and cost under $100. The beauty of USB is that you can plug in the video camera without cracking the computer case.

To make a video conference call, you use the software to dial the IP address of the party you're trying to reach. At the other end of the connection, the recipient's computer rings like a telephone or flashes a message on the screen alerting the participant of an incoming call. If the recipient accepts the call, each conferee appears on the other's screen, as shown in Figure 16-3. Chapter 18 explains adding video and voice to your Internet communications.

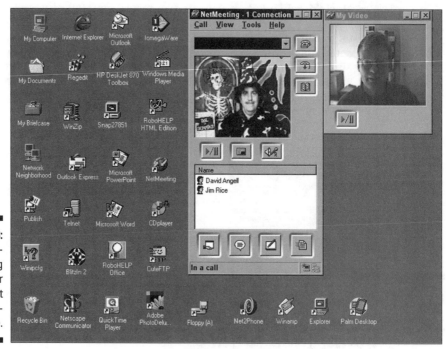

Figure 16-3:
Video conferencing brings your Internet communications to life.

Chapter 17

The Broadband-Powered Business

*Y*our Internet service is more than just a data communications link — it's becoming a vital connection to customers, employees, and business services. Time-honored business practices of delivering good customer service, improving employee productivity, and saving money are increasingly linked to how well your business can harness the power of the Internet.

The bottom line is that your business can no longer afford to be using outdated, slow, dial-up or ISDN Internet connections. This chapter takes you through what it means to be a broadband-powered business.

Going Broadband

The broadband-powered business uses the Internet as a natural extension of doing business. Over time, you'll find that the Internet gets used for an increasing number of activities, including the following:

✔ **Hunting and gathering information.** A high-speed, always-on connection makes surfing the Web for information, getting news, researching, and downloading software second nature.

✔ **Transacting business.** Your DSL connection gives you full-powered access to online business-to-business deals, financial management, and more.

✔ **Communicating.** Broadband powers video conferencing for keeping in touch visually with clients and colleagues. You can also harness your DSL connection to cut your voice communication costs using VoDSL (Voice over DSL). For more on VoDSL, see Chapter 18.

Experiencing Broadband Internet

Broadband means the Internet becomes a serious tool for your business. A DSL Internet connection delivers new and improved ways to use the Internet, such as the following:

✔ **Faster Web access.** This core function of DSL service makes the Web come alive, and the always-on feature of DSL means instant access. Faster Web surfing means your employees do more in less time.

✔ **Send and receive email instantly.** Your email is instantly sent — no waiting to dial up the Internet. You get mail whenever an email message is received at your email box. You can even run your own Internet email server for your business. This means you can operate a variety of email services, such as using an autoresponder for queries or automatic forwarding of email. See Chapter 20 for more information on running your own email server.

✔ **Move data and software quickly.** High-speed DSL means quick uploading and downloading of files. The more reliable digital nature of DSL service combined with its raw speed makes the movement of programs and files more feasible.

✔ **Connect multiple computers to a single DSL line.** Using a single, high-powered DSL connection allows you link all the computers in your business through a local area network.

✔ **Support telecommuters without modems and telephone lines.** Your business can add VPN (virtual private networking) capabilities to your DSL connection to allow secure connections between telecommuters, branch offices, and mobile users and your office. With VPN over DSL, you don't need banks of modems and telephone lines, so you can save a lot of money. See Chapter 19 for more information on telecommuting through VPN.

✔ **Collaborate remotely.** With the power of DSL, you can use video conferencing, voice, and collaboration software tools as a practical alternative to travel.

✔ **Do voice and video communications.** A DSL connection provides a great platform for integrating voice into your Internet communications as well as using Voice over DSL (VoDSL), which is a business-class voice service. With enough bandwidth, your business can even do video conferencing. See Chapter 18 for more information on DSL-powered video and voice communications.

Working Smarter with Broadband

Broadband affects your business from the minute your service is turned on. Like the ripples created in a still pond after dropping a pebble, the uses and benefits of DSL service expand in ever growing concentric circles.

As your business learns new ways to tap into all the Internet has to offer, your business works smarter. Your DSL-powered business gets a front row seat to all the new business innovations and solutions being delivered on the Internet today. Improved productivity takes hold as many of your business activities get converted into convenient online tasks. The case for powering your business with broadband is compelling.

Do more in less time

DSL is about doing more in less time. Faster, always-on Internet access through DSL means people are more productive. A PowerPoint presentation that would take an hour through a dial-up modem would now take minutes. You can research information from a number of Web sites simultaneously instead of waiting for each page to slowly appear one at a time. Email is instantly sent and received for quick turnarounds. Multiply all these timesaving benefits across all the people in your office and you see what speed means to your business.

Save money

A broadband connection can save you money on Internet connection costs. Even moderate use of a dial-up Internet connection can run more than $100 a month for the telephone line, business telephone usage costs, and Internet access. A small office with five PCs and dial-up Internet connections can easily incur costs of $500 a month. Converting to a 784-Kbps Internet connection can save your business up to 50 percent per month and deliver almost 20 times the speed of a modem.

Tap into Internet services

Online business transactions are about convenience, saving time, and saving money. Business-to-business services on the Internet are popping up like mushrooms after a spring rain. And your broadband business can take full advantage of them.

For example, you can use @Backup (www.atbackup.com) to make automatic backups of your critical data and store it offsite. No more tape backups, and no more taking the tapes home. You can create your own virtual Intranet using Intranets.com (www.intranets.com) or Webex (www.webex.com). All your essential network services are provided for you, including document management, group calendars, instant messaging, Web-based presentations, and more. No more maintaining your own expensive and complicated intranet.

Capture telecommuter productivity

Telecommuting isn't good for only employees and contractors; it's good for business as well. Allowing people to telecommute enables your business to share in their productivity gains. With fewer meetings, less interruptions, and no commute, productivity improves dramatically. Telecommuters routinely report productivity increases of 15 to 30 percent.

Broadband puts telecommuters in the fast lane for accessing your company network using VPN (virtual private networking). VPN is a way that private data can pass over a public network such as the Internet. It operates transparently over DSL, so telecommuters access resources on your network as if they're sitting in front of their computers at the office.

VPN isn't new. Big companies have been using it for years, but now affordable DSL makes it available to small businesses. Using VPN saves money on employee dial-up calls to your network. DSL-powered telecommuters and remote dial-up users use the Internet instead. Your business also eliminates the need for maintaining telephone lines and banks of modems to support dial-up connections.

One line does it all

High-speed DSL Internet service is made for sharing across all the computers in your business. No more separate telephone lines, modems, and Internet access accounts for each computer. Instead, you link all the computers in your business through a LAN (local area network). Don't have a network? Not a problem. Today, setting up a LAN is easy and inexpensive and you can choose from a variety of networking technologies to fit your needs.

Add visual communications

The broadband business can use video conferencing over the Internet to reduce travel time and costs. Although not the quality of a television image, real-time video conferencing adds valuable face-to-face contact with clients and colleagues. Your business can also use it as a customer service tool, to help telecommuters stay in the loop at the office, to enable remote expert and client support, to promote distance learning, and to enhance recruitment.

Make calls for less

VoDSL service uses your DSL connection as a conduit for voice calls to any telephone in the world. Your business can cram up to 16 "virtual" phone lines into a single DSL connection. With VoDSL, your business has new choices — beyond the local telephone company — in who provides your telephone service. You can get the same high-quality voice communications as regular telephone service but at a lower cost. With VoDSL, powerful voice-call-management features are available to small businesses at a fraction of the cost of expensive PBX (Private Branch Exchange) systems.

Your Broadband Application Strategy

The temptation to try out all kinds of cool, sophisticated business applications on your DSL connection is compelling. Stop, take a deep breath, and think. The rule of thumb about TCP/IP applications is: Just because you *can* do it doesn't mean you *should* do it. Each application has pros and cons that you need to look at carefully.

Here are some guidelines to keep in mind as you explore your TCP/IP application options and formulate your business strategy:

✔ Take a thoughtful strategy that compares the benefits to the commitment it takes to fully utilize an application. Trade-offs always exist between an application's benefits and costs.

✔ Be careful if you plan to run any type of server (such as a Web server or an email server) on your DSL connection. Running a public Web server on your LAN, for example, can make big demands on your DSL connection as well as on your time setting up and administering the server.

✔ A cool thing about DSL is that you can upgrade your speed to support growing demands without changing your DSL CPE or your IP address infrastructure. This means your services can evolve as you move up the TCP/IP application learning curve.

✔ Consider using a hosting service for your Web server, email, and other Internet services. Hosting a Web site and email boxes with your domain name is affordable, and many packages are available.

Chapter 18

Video and Voice over DSL

*T*he high speed and always-on features of the DSL connection makes using higher-order video and voice communication tools practical for consumers and businesses. This chapter introduces you to video and voice communications and how you can get started using these powerful tools.

The Next Best Thing to Being There

Video conferencing is the sexiest of the Internet applications. It brings exciting show-and-tell capabilities to your Internet communications. Broadband-powered businesses can use video conferencing to reduce travel time and costs or as a new customer service tool. Video conferencing can also help telecommuters stay in the loop at the office, enable remote expert and client support, promote distance learning and recruitment, and more. Broadband-powered households can use video conferencing to drop in on other broadband-powered friends and family.

DSL-powered video conferencing

Because video conferencing eats up bandwidth, you need the power of DSL to bring it to life. And because DSL is an always-on Internet connection, you can receive incoming video calls anytime your computer is turned on.

Video conferencing requires a DSL connection with enough upstream and downstream speed to support two-way video and audio transmissions. To get an acceptable quality video image, you should have at least a 384-Kbps DSL connection. If you're using ADSL (Asymmetrical DSL) or G.lite service, make sure the upstream speed (from your PC to the Internet) is enough to support the higher rate needed for video conferencing. Otherwise, the other participant sees a much poorer quality of your video.

How video conferencing works

With the video conferencing kit and software installed on your computer, you're ready to make and receive video calls. The Internet Protocol (IP) address assigned to your computer or the computer of the other participant acts as the unique telephone number.

To make a video conferencing call, you type — in the video conferencing software — the IP address (such as 199.232.255.113) of the party you're trying to reach. At the other end of the connection, the recipient's PC rings like a telephone or flashes a message on the screen, alerting the participant of an incoming call. If the recipient accepts the call, each conferee appears on the other's screen (see Figure 18-1).

Figure 18-1:
A video
conference
in action.

Most video conferencing kits are based on international communication and conferencing standards, including the International Telecommunication Union (ITU) H.323 standard for audio and video conferencing and the T.120 standard for multipoint data conferencing. The H.323 standard specifies the use of T.120 for data conferencing functionality, enabling audio, data, and video to be used together in a conference session. Support for these standards ensures that you can call, connect, and communicate with people using compatible conferencing products from other vendors. Other important standards that come into play with video conferencing are H.255.0, H.245, G.711, and H.261.

Dynamic IP addresses and video conferencing

Video conferencing uses an IP address as your telephone number. If you're using a DSL Internet connection in which the ISP assigns IP addresses dynamically, video conferencing gets more complicated. In a dynamic IP addressing connection, your IP address changes, so people making a video conference need to know your new IP address each time it changes. There are ways around this, however. For example, you can automatically logon to a video conferencing server on the Internet so people can reach you.

Firewalls, routers, and video conferencing

Video conferencing can run into problems when your Internet connection passes through firewalls or routers using NAT (Network Address Translation). Video conferencing software, such as CU-See-Me or Microsoft NetMeeting, uses dynamic TCP ports to deliver the data, and this isn't supported by certain firewalls, proxy servers, and routers. However, a growing number of DSL routers, proxy servers, and firewalls support using video conferencing.

What typically happens if the firewall or router isn't allowing the video conferencing data through? The other participant sees your video image but you don't see his or hers.

Lights, camera, action!

Video cameras are sensitive to different types of lighting. Take a little time to play with the lighting around you to improve the quality of your video image. Here are some basic video conferencing tips:

- ✔ Avoid strong back lighting.

- ✔ Use fluorescent or white incandescent lights, if possible. Clear incandescent lights can result in yellowish video.

- ✔ Avoid direct lighting from the front because it creates flat, unnatural looking images.

- ✔ Make sure your lighting is adequate. Too low a light creates grainy looking images.

- ✔ Dress for success by avoiding dark or white clothing, reds, or busy patterns, all of which negatively affect your image.

- ✔ Put the camera on your monitor with the recipient window directly under it at the top of your monitor screen. This allows your eyes to focus on the other participant.

Video Conferencing Kits

Today's video conferencing kits are amazingly affordable (under $125) and easy to set up. These kits include software and a small video camera that sits on top of your monitor. The software typically includes programs for video conferencing, creating Vmail (self-running video clips you send as email attachments), operating a WebCam, and more. Table 18-1 shows you the leading video conferencing kit vendors.

Table 18-1	Video Conferencing Products	
Vendor	*Web Site*	*Product(s)*
3Com	www.3com.com	3Com HomeConnect USB kit
Creative Labs	www.creativelabs.com	Video Blaster WebCam 3; USB video conferencing kit
Kodak	www.kodak.com	Kodak DVC323; USB video conferencing system

Vendor	Web Site	Product(s)
Winnov	www.winnov.com	Videum Conference Pro; PCI and USB video conferencing kit
Zoom Telephonics	www.zoomtel.com	ZoomCam; USB video conferencing kits for PC and Mac
Intel	www.intel.com	Intel Create & Share Camera Pack; USB video conferencing kit
LogiTech	www.logitech.com	LogiTech QuickCam Pro; USB video conferencing kit

Most video conferencing kits use USB (Universal Serial Bus) to connect the camera to a PC or a Mac. USB is the latest standard for connecting peripherals to a computer without cracking the case. USB ports have been standard on most PCs for the last few years. Microsoft Windows 98 Second Edition, Windows 2000, and MacOS (8.5 or higher) support USB.

Video conferencing is processor intensive, so the more processing power your computer has, the better. More memory and a faster display adapter also contribute to a better quality video image.

Keep your eye on the camera

The video resolution that the camera supports plays a big role in the quality of the image. A tradeoff always exists, however, between the quality of the image and the frames per second (fps) during any video conference. The higher the resolution, the lower the frames per second because high resolution generates more data to process and pass through a connection. There is also a trade-off between the size of your video conferencing window and the quality of your video image. The larger the window, the lower the video image quality.

Wired for sound

Video conferencing uses audio as part of the communications mix. Most video conferencing kits rely on your computer's sound card and speakers as well as a microphone to handle your voice communications. You need a full-duplex sound card, so that you can talk to and hear others at the same time. Most sound systems included with today's computers are full duplex.

The audio portion of a video conference can degrade the quality of your video image because it reduces the amount of bandwidth available for the video. You can offload the audio portion of a video conference to standard telephone service (preferably using a headset or a speakerphone) to improve the quality of your voice communications and free up bandwidth for video.

Take a look at the 3Com HomeConnect

3Com's HomeConnect video conferencing kit represents the state-of-the-art in video conferencing kits, for a street price of around $125. The HomeConnect video camera, shown in Figure 18-2, offers outstanding picture quality. The camera delivers resolutions from 128-by-96 pixels up to 1,280-by-960 pixels. The camera connects to a PC or Mac using USB. Software drivers for the PC are included on the CD; you can get the drivers for the Mac on the 3Com Web site.

Figure 18-2:
The 3Com Home-
Connect video camera.

The HomeConnect kit includes a complete bundle of software to take advantage of your video system, including

- ✔ PictureWorks Live to take video snap shots and video clip captures
- ✔ PictureWorks NetCard for creating Vmail
- ✔ Microsoft NetMeeting for video conferencing

Video Conferencing Software

With a video camera installed on your computer and the right software, you can do video conferencing, send images to your Web site using WebCam software, and create Vmail.

The two leading video conferencing programs are Microsoft NetMeeting and White Pine Software's CU-SeeMe. Both support the ITU H.323 standard for video conferencing and the ITU T.120 standard for multipoint data conferencing.

Microsoft NetMeeting

You can't beat the price for Microsoft NetMeeting — it's free. You can choose to download NetMeeting as part of the Internet Explorer package or you can download it separately from the Microsoft Web site at `www.microsoft.com/msdownloads/`.

Microsoft NetMeeting (see Figure 18-1) includes the six most wanted collaboration tools (audio, video, file transfer, chat, document/application sharing, and whiteboard). Video and audio are limited to only two conferees, but you can switch from one conferee to another on-the-fly.

CU-SeeMe

CU-SeeMe software isn't free, but it's also not expensive with a street price of less than $60. The performance and features of CU-SeeMe are comparable to what you get with Microsoft NetMeeting. CU-SeeMe includes some collaboration tools, such as a whiteboard, a file transfer utility, and a text-based chat option, but lacks document sharing.

One cool feature that CU-SeeMe has that NetMeeting doesn't is the capability to interact with up to 12 users simultaneously, with a video window for each. Figure 18-3 shows a multiuser conference at Cu-SeeMe World.

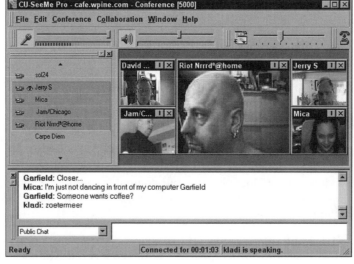

Figure 18-3:
A multiuser
video con-
ference
using CU-
SeeMe.

Creating a WebCam

A WebCam operates as a video surveillance camera pointed at just about any-thing you can image. You've probably run into Web sites with a WebCam. The WebCam software takes images from the video camera at designated inter-vals, such as every few seconds, and uploads them to any Web site you desig-nate. Web site visitors can then view the images, which are automatically refreshed in their Web browser.

The most popular WebCam software is WebCam32, which is bundled with some video conferencing kits. You can also buy WebCam32 at www.surveyor.com.

Sending Vmail

When you send Vmail, the video arrives as an email file attachment with the video and self-contained video player ready to play. The recipient simply double-clicks the file and there you are, in real time on their computer screen. The video and audio are compressed into a small file that you can send even to recipients who use a modem connection to the Internet.

Some popular Vmail programs are

- ✔ NetCard from PictureWorks (www.pcitureworks.com)
- ✔ Vmail from SmithMicro (www.smithmicro.com)
- ✔ VDOMail from VDO (www.vdo.net)

Let's Talk about Voice Communications

A DSL connection provides a great platform for integrating voice into your Internet communications. You can choose from three technologies to do your talking over a DSL connection:

- ✔ **Computer to computer through the Internet.** Products such as Microsoft NetMeeting, Net2Phone, and Internet Phone allow you to do voice communications using the sound card, speakers, and microphone (or headset) connected to your computer.

- ✔ **Computer to telephone through the Internet and PSTN.** This form of IP voice communication uses a hybrid system of Internet-based voice that's passed over to the PSTN to connect to a telephone.

- ✔ **Voice over DSL (VoDSL).** A hybrid system that includes voice over IP or voice over ATM between the customer and the DSL provider network, which is passed on to a VoDSL voice provider and then passed on to the PSTN.

VoIP forms the basis of voice communications over the Internet. VoIP also forms the basis of VoDSL but instead of using the Internet, this VoIP uses the DSL connection from only your premises to the CO before passing it over to a private IP-based network or the PSTN.

Doing voice over the Internet

Net Phone is VoIP software that sends your packetized voice over the Internet. These packets are mixed in with all the other data traffic. If Internet traffic is heavy or the route your data travels suffers from bottlenecks, you experience delays in conversation — the person on the other end won't hear your sentence until all the packets have arrived. When packets don't arrive, you experience dropouts, or missing words or phrases. The big appeal of

VoIP using the Internet is that the calls are free. This makes voice over the Internet popular for making otherwise-expensive international calls.

You can also use VoIP as a way to make voice calls from your computer to a telephone. Your voice call goes over the Internet until it reaches a VoIP-to-PSTN switch, which converts the VoIP call to the PSTN to ring the telephone at the other end. This hybrid service is usage-based but offers big savings on usage charges. Net2Phone and others offer this type of service. Figure 18-4 shows the Net2Phone software used for making voice calls.

Figure 18-4:
The
Net2Phone
software for
making PC-
to-PC and
PC-to-phone
calls.

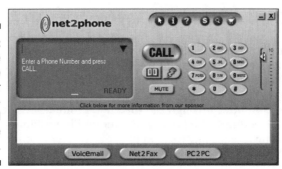

Net Phone software combined with a sound system on your computer allow you to make and receive voice calls over the Internet and also reach out to people with voice telephones through the PSTN network. To make calls from the Internet to a telephone, you need to use one of the Internet-to-PSTN services.

Here are the leading PC-to-PC and PC-to-phone software and service providers:

✔ Internet Phone (www.vocaltec.com)

✔ Net2Phone (www.net2phone.com)

✔ PhoneFree (www.phonefree.com)

Voice over DSL

Voice over Digital Subscriber Line (VoDSL) doesn't use the Internet; it uses the DSL network and then passes the voice over to the PSTN. As a result, VoDSL communications have the same high quality as regular telephone service. VoDSL is new and was not widely deployed as of this book's writing, but you can expect to see it arriving at a DSL connection near you in late 2000.

New VoDSL technologies allow you to cram up to sixteen phone lines onto one by splicing the digital spectrum of the DSL connection. This enables a small business, for example, to have five voice connections and fast Internet access all from the single DSL line.

Why VoDSL?

VoDSL opens up access to competing voice carriers beyond the local telephone company. VoDSL, with prioritization of the voice traffic, enables businesses to use a single line to connect both their voice and data traffic to the carrier's core network — while preserving toll-quality voice.

How does VoDSL work?

VoDSL uses your DSL connection as a conduit for your voice communications. Your voice is packetized for transport over the DSL network and then handed off to a voice gateway. Unlike packetized voice over the Internet, VoDSL creates a priority channel in your DSL connection to handle the voice communication. A voice gateway at the CO (or other location on the DSL network) unpacketizes the voice and hands it off to the PSTN to be routed as any ordinary voice call. Figure 18-5 shows how VoDSL works.

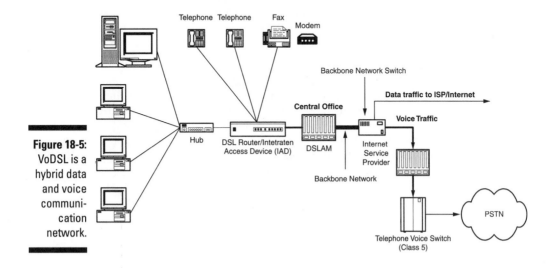

Figure 18-5: VoDSL is a hybrid data and voice communication network.

The leading VoDSL gateway makers include:

- CopperCom (www.coppercom.com)
- JetStream (www.jetstream.com)
- TollBridge (www.tollbridge.com)

An IAD (Integrated Access Device) at your premises packetizes your voice going out over the DSL network. You plug in standard analog telephones, fax machines, and modems into the IAD. The IAD can be either a standalone device or integrated into a DSL router, such as the Efficient Networks SpeedStream 7451. (The SpeedStream is a DSL router with four VoDSL ports for connecting analog telephones, fax machines, and modems to the DSL connection.)

When you make a call, 64 Kbps is taken out of your data service and used for an IP voice link. You dial your telephone number as you would for any telephone call. The voice gateway, which resides in the service provider's network, receives traffic from the IAD in packet format, usually by way of a DSLAM at the CO. It reconstructs the call to pass over to the PSTN voice switch, which is called a Class 5 switch.

What is toll quality voice? Telephone voice quality is achieved with a maximum one-way delay of up to 50 milliseconds. At this rate, the delay is not perceptible to the human ear. Voice over the Internet can add higher delays that affect the quality of your conversation.

Getting VoDSL service

Voice CLECs are partnering with data CLECs to deliver voice and data communications over a single DSL connection. Current partnerships in the VoDSL arena include Rhythms NetConnections with MCI WorldCom, Covad Communications with GST Telecommunications, and NorthPoint Communications with Focal Communications. Chances are you'll get your VoDSL service though an ISP reselling a voice CLECs service in much the same way they resell a data CLEC's DSL service.

Chapter 19

Telecommuting in the Fast Lane

*A*n undeniable combination of business and social trends is driving telecommuting. In today's marketplace, businesses find that if they want to keep their best employees, they must answer their demands for more flexible work schedules. In addition, some businesses want to retain talent beyond the local market, without investing in huge relocation fees. For both issues, telecommuting provides businesses and workers alike with the solution.

What's Virtual Private Networking?

Virtual Private Networking (VPN) is a way that private data can pass over a public network such as the Internet. A VPN can connect individual telecommuters to the office network, creating a separate tunnel for each connection, as shown in Figure 19-1. A VPN can also connect remote office networks together as a LAN-to-LAN connection over the Internet using a single tunnel, as shown in Figure 19-2.

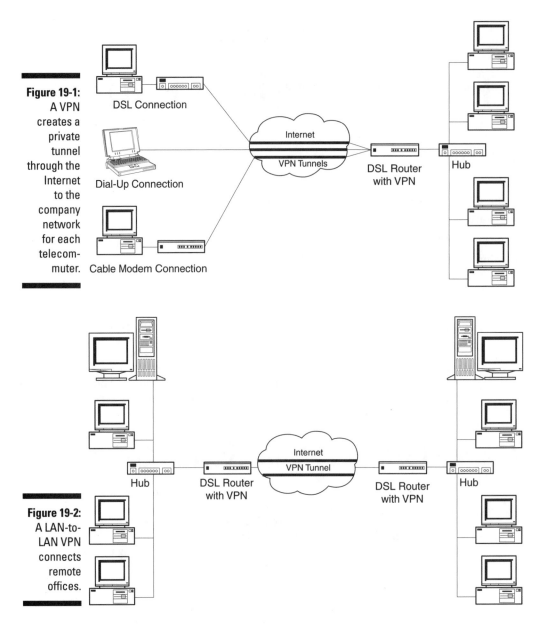

Figure 19-1: A VPN creates a private tunnel through the Internet to the company network for each telecommuter.

Figure 19-2: A LAN-to-LAN VPN connects remote offices.

From the telecommuter's perspective, the VPN is a point-to-point connection between the user's computer and the company server. The telecommuter uses a VPN client on a computer at home that connects to the Internet, which connects to a VPN server at the office. The VPN connection operates transparently over DSL, so the telecommuter works on the office network as if he or she was sitting at a computer in the office. As the data passes between the telecommuter and the company network, it's encrypted to keep it private as it passes through the Internet.

A few small business VPN solutions

VPN has traditionally been the province of large enterprises, which use VPN to replace costly private networks linking remote sites with less expensive Internet connections. The cost of these industrial-strength VPNs, with prices ranging from $4995 to $9995 and up, are out of reach for smaller businesses.

With the advent of DSL, small businesses now have access to the dedicated Internet service that can support a VPN. A few small business VPN solutions are available today, but you can expect to see more products soon.

Where to get VPN

Expect to see VPN solutions packaged and priced for nontechnical small businesses as DSL-powered businesses demand support for telecommuters. Creating a VPN solution for small businesses can be accomplished at the network edge using DSL routers with VPN server functionality or third-party hardware and software products. The VPN service can also be performed at the network level of the DSL provider or ISP.

If you want to get VPN installed, check out TechPlanet by visiting `www.tech planet.com` or calling 877-832-4752. TechPlanet is a nationwide technology service company that does on-site VPN, network, and other technology installations and provides support specifically for small businesses.

I Want My VPN

In today's competitive business climate, companies face demands for more flexible work schedules. To attract and retain the best and brightest people, businesses need to support employees who want or need to telecommute or work extended hours outside the office. Telecommuting allows a business to retain talent beyond the local market. According to IDC/Link, 65 percent of telecommuters work for companies with fewer than 100 employees.

More than 14 million people will be telecommuting by the year 2000, according to research company FIND/SVP, and small business employees make up the bulk of those telecommuters.

A flexible work environment is good not only for employees but also for business. With fewer meetings, less interruptions, and no commute, productivity improves dramatically. Telecommuters routinely report productivity increases of 15 to 30 percent. With the advent of DSL, telecommuters are even more productive because the high-speed connection makes them work faster on the company network.

DSL is the diamond lane of telecommuting

Using a DSL connection, telecommuters can work more productively at home and get the added bonus of capturing lost time from commuting on congested highways. DSL-powered telecommuters work smarter from home because they have access through VPN to all the essential services of the company network, including email, databases, files, and Intranet resources. DSL also enables telecommuters to stay in the loop at the office through the use sophisticated tools, such as Web-based presentations and video conferencing.

How your business benefits from VPN

The marriage of DSL and VPN delivers some compelling benefits for small businesses, including the following:

- ✔ **Save telephone charges.** Remote users bypass the telephone system for long-distance dial-up calls. Instead, DSL-powered telecommuters connect through their flat-rate DSL connection. Remote dial-up users can use local Internet access numbers for connecting to the company network instead of making long-distance calls.

- ✔ **Bypass modem banks.** Connecting remote users through the Internet means that your company doesn't have to purchase and manage banks of modems with remote access server software to support dial-in users. Instead, remote access users connect to your LAN through the same Internet connection your company uses to connect to the LAN. All data traffic comes through the DSL connection.

- ✔ **Add security for remote access connections.** VPN enables you to secure your telecommuter connections to protect your confidential data.

- ✔ **Improve telecommuter productivity.** High-speed VPN means telecommuters can do more in less time from their home computers.

Inside VPN

VPN operates at Layer 2 and Layer 3 of the TCP/IP stack. The TCP/IP protocol has five layers: physical (layer 1), data link (layer 2), Internet (layer 3), transport (layer 4), and application (layer 5) layers. VPN technologies operate at the data link and Internet layers.

Why do you care what layer a VPN operates on? The difference of a VPN operating at layer 2 versus layer 3 affects the way the remote network looks and feels to clients, as follows:

- ✔ **Layer 2,** the data link layer, supports multiple networking protocols, such as NetBEUI (Microsoft's networking protocol). This means you can connect a VPN client to a Windows computer and use the Windows interface to work on the office network. For example, you'd navigate files using Windows Explorer.
- ✔ **Layer 3** supports only a single network protocol, such as TCP/IP. This means VPN clients interface with your network through office servers that consist of Web servers, FTP servers, and other TCP/IP applications.

Three protocols are used for today's VPN solutions: PPTP (Point-to-Point Tunneling Protocol, L2TP (Layer 2 Tunneling Protocol), and IPSec (IP Security). PPTP and L2TP are layer 2 VPN protocols; IPSec is a layer 3 VPN protocol.

PPTP and L2TP are better suited for client-initiated tunnels, and IPSec is best for LAN-to-LAN tunnels. PPTP and L2TP are more suitable for use in multiprotocol non-IP environments, such as NetBEUI; IPSec is designed to handle only IP packets.

These VPN technologies are packaged in different ways. Some VPN solutions are built into a DSL router; others are embodied in a standalone firewall/VPN security appliance that sits between the DSL bridge and the LAN (hub).

PPTP (Point-to-Point Tunneling Protocol)

PPTP is included with Microsoft Windows 95, 98, and NT. The VPN client software is included with Microsoft Windows as part of Dial-Up Networking (although it can also be used over a dedicated connection). The PPTP server is part of Remote Access Service (RAS) and Routing and Remote Access Server (RRAS), which runs on Windows NT Server and uses existing Windows user domains for authentication.

PPTP is a layer 2 protocol, so it supports IP, NetBEUI, and IPX. The support for NetBEUI, the Microsoft networking protocol, means small businesses can use an existing Windows network infrastructure instead of changing to IP client/server solutions. PPTP is the least robust of the VPN solutions in terms of security, but it's also the easiest to set up and the least expensive to deploy. PPTP does not include data encryption.

L2TP (Layer 2 Tunneling Protocol)

L2TP is the successor to PPTP. It combines many of the features defined in PPTP with those created for Cisco's Layer2 Forwarding (L2F) protocol. Because L2TP is a layer 2 protocol, it offers the same flexibility as PPTP for handling multiple protocols (IP, IPX, and NetBEUI). Microsoft Windows 2000 includes support for L2TP. L2TP is a more robust VPN solution than PPTP but is also more complex and expensive to implement.

IPSec (Internet Protocol Security)

IPSec is part of the next generation of Internet Protocol, IPv6. This IP security standard covers authentication and encryption of IP traffic between routers, host to host, or any user to host. IPSec is a standard-in-progress, however, which means different vendor implementations might not work with others.

To handle remote users, small businesses that want to use IPSec must have a LAN network with IP application servers. This means a business needs an Intranet to support telecommuters. IPSec is the most robust of the VPN technologies.

A DSL VPN Sampler

Some VPN solutions are available for small businesses. Here is a sampling of three VPN solutions. Two use VPN server software running on a DSL router; the third uses a VPN server that is part of a firewall product. The client software for these VPN solutions includes the Microsoft VPN client or third-party VPN client software.

VPN Lite using PPTP

You can create a VPN solution easily using PPTP (Point-to-Point Tunneling Protocol) with a DSL router or Windows NT/2000 acting as the PPTP server. Using a DSL router running as a PPTP server means that your LAN doesn't need to have Windows NT/2000 Server running.

The benefit of using PPTP is that the Microsoft VPN client is available at no cost because it's included with Windows 95/98/NT/2000. For a small business not running Windows NT/2000 Server, using a DSL router as a PPTP server, such as the Netopia R7100, makes setting up a VPN Lite solution inexpensive. Clients can connect to the Netopia R7100 using a dial-up, DSL, or cable Internet connection. Because of the added demands from PPTP processing on the Netopia router, this solution is not recommended for more than five simultaneous VPN connections.

Midrange L2TP solution

Your can create a more sophisticated VPN using the L2TP (Layer 2 Tunneling Protocol) server add-on for the FlowPoint (recently acquired by Efficient Networks) 2200 SDSL router. The list price of this upgrade is $195. A small business can choose an L2TP VPN to support individual connections using client software on remote PCs or as a LAN-to-LAN VPN using a FlowPoint router at each end of the connection. FlowPoint also includes a sophisticated firewall with its 2200 SDSL router at no additional charge, which enables a business to protect its LAN from Internet intruders.

FlowPoint uses an L2TP client called WinVPN from iVasion (recently acquired by WindRiver Systems). This client sells for around $70 each. Windows 2000 will also be supporting L2TP in its built-in VPN client. Configuring the FlowPoint router for VPN requires command-line configuration, which makes self-configuration by the nontechnical small business unlikely.

A basic IPSec VPN

The SonicWALL SOHO security appliance can be upgraded to support IPSec-based VPN for a total list price of $795. This SonicWALL VPN is a box-to-box solution designed for use between two small offices that have a SonicWALL device at each end of the connection. This creates a single tunnel between the two LANs for handling all the data traffic between them.

This site-to-site approach is recommended because the demands of multiple VPN client connections to a single SonicWALL SOHO puts too much processing overhead on the unit's processor. The SonicWALL VPN solution also includes a sophisticated firewall, and its Web interface makes it one of the easiest VPN solutions to set up.

An industrial strength VPN

The SonicWALL Pro and Netopia S9500 offer a more serious firewall and VPN solution. They both use IPSec as the VPN protocol and include dual processors to handle the demands of a firewall and a VPN. Both are security appliances that sit between a DSL bridge or router and the LAN (hub). They're good solutions for businesses that take security and remote access seriously. The downside of these products is that they're expensive, with a list price of $2995 plus the cost of client software at $70 a pop.

Remote Control Software as a VPN

Remote control software enables a remote client computer to take control of another computer (called a *host*) through an Internet connection. The remote client takes full control of the host computer system as if the remote user were sitting at the host computer.

Remote control over the Internet through TCP/IP allows you to access office computers from a local ISP connection anywhere on the planet. Because DSL is an always-on server, you can use remote control software to get access to a computer running the remote control software anytime.

The leading remote control software programs include the following:

- ✔ Symantec's pcAnywhere (www.symantec.com)
- ✔ Netopia's Timbuktu Pro (www.netopia.com)
- ✔ LapLink 2000 (www.laplink.com)

You can use remote control programs as a low-cost VPN solution to support remote access between two computers. To use remote control software, the host computer must be accessible from the Internet through a routable Internet IP address.

Part V
The Part of Tens

The 5th Wave   By Rich Tennant

"BETTER CALL MIS AND TELL THEM ONE OF OUR NETWORKS HAS GONE BAD."

In this part . . .

As its name implies, the Part of Tens imparts tens upon tens of valuable resource nuggets to help you on your way toward DSL enlightenment. In this part, you find helpful information on how to shop for the best DSL service and ten important questions to ask about DSL service.

Chapter 20

Ten DSL Service Shopping Tips

In This Chapter

▶ Compare DSL service offerings

▶ Find out whether you can order online

▶ Understand what you want from a DSL Internet connection

▶ Know your CPE options and prices

▶ Identify which DSL flavors they're selling

▶ Find out whether you can share the DSL connection

▶ Avoid a long-term commitment

▶ Get a written quote

▶ Read that fine print

▶ Determine the real speed

*T*he DSL service terrain is a maze of service offerings in all sizes and shapes. To tap into the promise of high-speed, always-on Internet connectivity, you need to be an informed DSL consumer. What you don't know can cost you. This chapter gives you some pointers to help you get the best DSL service for your needs and budget.

Shop Around

The Telecommunications Act of 1996 has unleashed a lot of competition in DSL. Use it to your advantage. Chances are your telephone company is offering DSL in your area — but don't stop there. Check out data CLECs as well. In most large metropolitan areas, you have choices from multiple CLEC DSL providers. Different DSL providers offer different service configurations and pricing options through their ISP partners.

Comparative shopping for DSL service is a complex endeavor that requires looking beyond just the price. You need to compare apples to apples, not apples to oranges. Low prices are great but don't assume that all service is created equal. You have to ask questions to find out exactly what you're getting for your money. What are the upstream and downstream speeds? What type of DSL CPE do they use? Do they offer static or dynamic IP addresses? Are there any restrictions?

If you're in an area where both CLECs and ILECs offer DSL service, you have access to a variety of different service options. You can shop for DSL from different types of ISPs:

- ✔ Independent ISPs use DSL service from multiple DSL providers.
- ✔ ILEC ISPs offer Internet access services using only the DSL offerings from the ILEC.
- ✔ ISPs acting as CLECs are data CLECs providing their own DSL service.

Many DSL providers and ISPs offer special deals that can save you some cash. For example, an ISP might offer free installation or a discount on DSL CPE. Many times, these promotions are offered by the DSL provider and passed on to the customer through the ISP. Ask for any deals the ISP might be offering and check ISP sites for any special DSL service specials.

See Whether You Can Order Online

Your time is valuable, so you don't want to be listening to music while you're on hold waiting to talk to an ISP salesperson. Check the ISP Web site to see whether it has all the information you need to make a decision. If you decide to go with the ISP, see whether you can order online. If you can't order the service at the ISP Web site, check out the DSL provider's Web site. Many offer online ordering but be careful — they might steer you to an ISP that doesn't offer you the best deal.

Know What You Need

Before you make the call to the ISP to start talking turkey about their DSL service, you should have a good idea of what elements you want as part of your DSL service. You need to ask yourself a bunch of questions, including the following:

- ✔ What TCP/IP applications do I want to use?
- ✔ What IP address configuration do I want to use?
- ✔ What are my current and future bandwidth needs?
- ✔ What DSL flavors are available in my area?

  ✔ Will I want to connect more than one computer now or somewhere down
    the road?

  ✔ Do I want bridged or routed service?

  ✔ What are my security options?

Know Your CPE

DSL CPE plays an integral role in the DSL service package because it defines
what you can and can't do with your DSL connection. For example, a DSL
provider might offer a high-speed DSL Internet access package that looks
great — until you discover that the DSL CPE they're using restricts the service to a single computer. Although you can get around this limitation, doing
so involves buying and setting up more hardware or software.

Remember, the differences in CPE technologies — such as PCI adapter, USB
DSL modem, Ethernet bridge, and router — have a big effect on your DSL service. Carefully review the CPE options offered by the ISP. Different ISPs might
support only a few of the CPE products that work with a particular DSL
provider's DSLAM.

Most ISPs keep CPE prices low because they don't want to discourage new
DSL customers with high start-up costs. But CPE prices do vary among ISPs,
so compare.

Determine Which DSL Flavors They Sell

Different DSL flavors (ADSL, SDSL, G.lite, and IDSL) each have there own
inherent characteristics. All deliver higher speed, but make sure you know
the upstream and downstream speeds for any ADSL service package. Some
ILECs offer very slow upstream speeds with their service, such as 90 Kbps to
128 Kbps, which restricts the real benefit of DSL for two-way communications
(video and audio), file uploads, and large email attachments.

A growing number of ADSL and G.lite DSL consumer offerings are using PPP
over ATM (PPPoA) or PPP over Ethernet (PPPoE). Using this technology creates a dial-up connection over an always-on DSL line. What it's actually doing
is authenticating to connect to the ISP's server — the way you do for a dial-
up Internet account. PPPoA forces you to use the same dial-up properties in
Windows or the Mac that you use for an analog or ISDN modem, even though
DSL is an always-on connection. In addition, a DSL modem using PPPoA can
be klugey because it installs as a network adapter but uses Microsoft
Windows Dial-Up Networking to make the Internet connection.

PPPoA and PPPoE have other drawbacks for DSL subscribers. As a consumer DSL offering, they promise to be heavily oversubscribed. DSL providers and ISPs can easily oversubscribe PPPoA connections like they do for dial-up connections. The result is that you might not be able to connect to the Internet because all circuits are busy.

Find Out Whether You Can Share the DSL Service

Any DSL service you're considering should be weighed against whether you think there's a chance that you'll want to connect another computer to the DSL line. Many low-cost consumer DSL offerings have restrictions against sharing the DSL connection across multiple computers. They use a PCI DSL adapter card or a USB modem (which are more difficult to share across a network) or have restrictions against sharing a connection using NAT.

The best type of consumer DSL service in terms of being able to share it easily is the one that uses an Ethernet modem with at least one static IP address.

Read the ISP's Terms and Conditions contract before setting up any kind of server. Here's an excerpt from the Terms and Conditions (January 2000) document for Flashcom's SoloSurfer service regarding server restrictions:

> Restrictions. SoloSurfer and SoloSurfer Express Customers agree not to run any servers in conjunction with the Services, including but not limited to, electronic mail, NAT, DHCP, and DNS servers. In the event any SoloSurfer or SoloSurfer Express Customer attempts to utilize a server on the network, Flashcom may, at its sole discretion, increase fees associated with the Services, or terminate the Services.

Most consumer DSL service offerings from other ISPs have similar restrictions. Server restrictions are usually removed for business DSL service offerings, but check the Terms and Conditions of any DSL offering to make sure.

Chapter 10 explains options for sharing an Internet connection across multiple computers.

Avoid a Long-Term Commitment

Many DSL service packages are contracted for anywhere from one to three years, with penalties for early termination. Although the ISP might offer a better price break for a long-term contract, be aware of the inherent dangers with long-term commitments for DSL service. The biggest problem with a long-term commitment for DSL service at a fixed price is that DSL service pricing trends are generally downward. If prices drop for DSL service, do you want to be stuck paying the older, higher rate? Ask the ISP for a price protection deal that allows your DSL price to go down if they lower their prices.

Get a Written Quote

Shopping for DSL for a business means picking up the telephone and calling the ISP to work out your DSL service package. After you've discussed what you want and what it will cost, get a written quote to confirm your verbal discussion. An ISP's quote should provide the following information:

✔ The DSL service and speed you specified

✔ The CPE used for the service and its cost

✔ The breakdown of one-time set-up charges

✔ Total monthly recurring charges

✔ Additional charges for custom services, IP addresses domain services, and any additional email boxes

✔ A copy of the terms and conditions of your DSL service

Read the Fine Print

Most ISPs have a terms and conditions document that spells out what you can and can't do with your DSL service. These contracts are important and you should read them carefully. The terms and conditions in this DSL service contract can make or break a DSL service deal. Here are some important things to look for in a DSL service contract:

✔ **What is the time commitment for your DSL service?** Many ISPs require that you make at least a one-year commitment to the DSL service, with some ISPs requiring a two- or three-year commitment. Depending on the provider, the cost per month might drop if you commit for a longer period. Before committing to a long-term contract, consider that prices will probably come down as competition heats up. Also check for early termination charges.

✔ **Are there usage restrictions?** Many ISPs include restrictions on the amount of data going through your DSL pipe per month per line. For example, there might be a 10-gigabyte limit for a lower-bandwidth DSL connection. Any data moving across your DSL line in excess of the limit costs extra, usually based on a price per megabyte. You might find an even harsher restriction that forbids you from running any servers or using a router that supports NAT and DHCP to share the connection.

✔ **What are the terms of payment?** Increasingly, many ISPs are billing to credit cards to cut down on their accounts receivable overhead. Others will bill you on a monthly or annual basis, with the annual billing method offering a discount. Many ISPs also require a deposit to start the service.

Find Out the Real Speed

Most DSL service is sold as a best-effort service, which means that the DSL provider and ISP do what they consider to be their best to deliver high-speed Internet access. The speed being touted by DSL providers and ISPs is really the speed supported by the DSL line from your premises through the DSLAM at the CO. The DSL network also includes backbone network connections from the CO to the DSL provider's network center and then on to other network connections to the ISP. From the ISP, the data goes out to the Internet through another network connection.

DSL providers and ISPs are in the business of leveraging available bandwidth to get the most bang for the buck. The urge to oversubscribe is a compelling economic issue and they all play the bandwidth game. Oversubscribing could become intense with DSL deployment because ISPs offering relatively inexpensive high-speed connections to customers might seek to cut corners on their backbone connections.

 As a potential DSL customer, finding out your real-world speed is almost impossible until you get the service. Your only real defense is to research what others are saying about the service from different DSL providers and ISPs. One helpful site for finding out what DSL subscribers think about different DSL providers and ISPs is DSL Reports at `www.dslreports.com`.

Chapter 21

Ten Questions to Ask about DSL Internet Service

....

In This Chapter

▶ Finding out about the availability of DSL

▶ Defining your need for speed

▶ Determining the type of CPE you want to use

▶ Uncovering your IP configuration

▶ Computing the cost of your DSL service

▶ Finding a helpful ISP Web site

▶ Unearthing terms and restrictions

▶ Checking on how an ISP/DSL provider's service stacks up

▶ Connecting more than one computer

▶ Thinking about Internet security

....

*T*he process of establishing a DSL connection involves evaluating interrelated components that together make up your complete DSL service package. This chapter contains the key questions you'll want answered as part of checking out your DSL Internet service options.

Is DSL Service Available in Your Area?

You can find out about DSL service availability from ILEC, CLEC, ISP, and other Web sites. Chapter 4 provides a detailed listing of the major ILECs and CLECs offering DSL service and their ISP partners.

You can also use Web sites such as www.dslreports.com and www.getspeed.com to check out DSL service availability. The www.dslreports.com site also provides database-driven user feedback rankings of ISPs offering DSL service.

What's Your Need for Speed?

In the real world, the biggest constraint on bandwidth is cost. The more bandwidth you want, the more it will cost. Bandwidth capacity planning is one of the most important and most difficult tasks you'll undertake in setting up a DSL connection. A number of factors come into play when trying to evaluate your bandwidth needs.

Here are some basic bandwidth questions you need to answer:

- ✔ **Will you be running any servers on your DSL connection?** A big factor in determining your bandwidth requirements is whether you plan to run any servers on your DSL connection. If you plan to run a Web server, for example, you need to take into consideration the incoming traffic as well as your outgoing traffic.

- ✔ **Do you plan to use IP voice and video conferencing?** If used even moderately, these cool but bandwidth-hungry applications can eat up your DSL connection. Supporting multiple simultaneous users compounds the demands.

- ✔ **Do you plan to connect one computer or a network to your DSL line?** The more PCs sharing the DSL connection, the greater the demand for bandwidth.

- ✔ **Can you upgrade to a higher speed?** One of the best features of many DSL service offerings is their bandwidth *scalability*. This means you can upgrade to a higher speed over time without having to start over again with new CPE.

Which Type of CPE Do You Want to Use?

What stands between you and the Internet through DSL is the Customer Premises Equipment (CPE). The type of DSL CPE you use as part of your DSL service plays an integral role in defining your Internet connection capabilities. Because DSL offerings are typically sold as a complete package that includes the DSL service, Internet access, and CPE, you need to understand the differences in DSL CPE offered as part of a package. Related to your CPE decision are IP configuration considerations, which are explained next in the "What's Your IP Configuration?" section.

DSL CPE devices break down into single-user or multiuser solutions. Single-computer DSL CPE consists of the following:

✔ PCI (Peripheral Component Interconnect) adapter cards

✔ USB (Universal Serial Bus) modems

✔ Ethernet bridges

These single-user solutions are often bundled with dynamic IP Internet access accounts from the ISP. This means that the IP address changes depending on the lease times established by the ISP. Most dynamic IP Internet service doesn't support domain name service, which means you can't use your own domain name as part of the service. Many ILEC, ADSL, and G.lite offerings from in-house ISPs use dynamic IP addressing.

Dynamic IP Internet access is usually bundled with CPE that restricts access to only a single computer. Most of these single-user DSL CPE options can be modified to support multiple users by using proxy server software, Internet-connection-sharing software, or Ethernet routers. Proxy server software and Ethernet routers also provide firewall protection for your LAN. Chapter 9 goes into detail on proxy servers and Internet security.

DSL CPE for the multiuser network environment consists of the following:

✔ LAN modems (Ethernet bridges)

✔ Routers

DSL bridges are often referred to as DSL LAN modems. A DSL LAN modem connects to a hub, which makes the DSL service available to all computers connected to the hub or switch. Bridges are simpler internetworking devices that move all TCP/IP traffic without filtering or routing the data.

A router is a more sophisticated gateway device than a bridge. A router allows data to be routed to different networks based not on hardware addresses (as in a bridge) but on packet address and protocol information. This decision-making functionality, called *filtering,* not only enables a router to protect your network from unwanted intrusion but also prevents selected local network traffic from leaving your LAN through the router. This is a powerful feature for managing incoming and outgoing data for your site. Because of the enhanced features included in routers, they cost more than DSL modems (bridges), often by a few hundred dollars.

See Chapter 8 for more information on all DSL CPE options.

What's Your IP Configuration?

The type of IP addressing you use as part of your DSL Internet service plays a pivotal role in determining the kind of interaction you have with the Internet. IP addressing defines what you can and can't do with your Internet connection. A dynamic IP address configuration is targeted to consumers who don't plan to run TCP/IP applications that require a static IP address. You can share a dynamic IP Internet account using a router with NAT, proxy server software, or an Ethernet router.

Using static IP addresses means using IP addresses that are recognized and routable on the Internet. Businesses and power-users typically use this type of IP addressing. Static IP addresses can be linked to specific hosts and domain names and enable Internet users to access a host computer running as a Web server (or any TCP/IP application server) using a user-friendly text identifier, such as www.angell.com. You can get consumer DSL service that uses static IP addressing, but in most cases the ISP won't let you use DNS service.

Routed, DSL service requires three IP addresses just for the IP server. One IP address is for the router, one is for the Ethernet connection, and one is for the WAN connection. If you get a block of eight IP addresses, for example, only five are available for hosts on your LAN.

Blocks of IP addresses are available from most ISPs for a monthly cost based on the number of addresses. Some ISPs include a block of IP addresses as part of the service. The IP addresses assigned to you by the ISP are available for use while you are the ISP's DSL customer. They remain the property of the ISP and return to the ISP upon termination of the service.

The ISP also provides name resolution service so that TCP/IP applications can use Internet domain names instead of just numeric IP addresses. The ISP will typically register a domain name on your behalf for no cost (although you can reserve the domain name anytime before signing up for your DSL service), but you will be billed for the domain name directly from the registrar. If your domain was previously hosted at another ISP or hosting service and you want to move it to a new ISP, the new ISP will typically do it without charge.

Chapter 7 deals with TCP/IP considerations as part of your DSL service package.

What Are the Total DSL Charges?

The total price you pay for your DSL service depends on the type of service you selected. You might have a one-time charge for getting started that includes the installation and CPE. Many DSL consumer packages include free (or near free) installation and a DSL modem for a one-year commitment. After your service is up and running, you have a recurring monthly charge for the DSL Internet service. You might have additional monthly charges for extra ISP services, such as more email boxes or IP addresses.

How Helpful Is the ISP's Web Site?

The ISP's Web site is an important starting point for your DSL service shopping. A good Web site should help customers develop their package before they call to talk to the ISP salesperson. Unfortunately, many ISPs lack good DSL customer information at their sites.

Here are some guidelines to judge a helpful ISP Web site:

- **Does the Web site provide specifics on the ISP's DSL service offerings?** Glossy marketing copy doesn't make for good consumer information. Information packaged to educate the consumer about the service and product details is helpful.

- **Does the Web site include the costs of the ISP's DSL offerings?** Installing and using DSL service involves a variety of charges. A Web site should provide a breakdown of the costs for getting the DSL service, including a menu of optional services.

- **Does the Web site include the ISP's terms of the service?** Unfortunately, most ISPs don't post their terms and conditions on their Web sites. These documents are the fine print of your DSL service. Terms and conditions are usually provided as part of the formal quote.

- **Does the Web site provide good CPE information?** CPE is at the heart of your DSL service capabilities and defines your TCP/IP application options. The Web site should provide coverage of the different CPE options and their prices.

- **Can you order online?** After you get the facts and decide you want to order, does the ISP offer you the convenience of ordering online?

- **Does the Web site provide information on IP addresses and domain name services?** A good Web site lists your IP address options and costs, such as the additional costs for IP addresses and DNS registration. The site should also include a menu of custom services, such as ISP support for running your own email server.

What Are the ISP's Terms and Restrictions?

The terms and conditions in the DSL service contract can often make or break a DSL service deal. Restrictions on your DSL service are spelled out in this contract, so you must read it carefully to fully understand what you can and can't do with your DSL service.

Here are some important things to look for in a DSL service contract:

- **What is the time commitment for your DSL service?** Many ISPs require that you make a one- to three-year commitment to the DSL service. With some providers, the cost per month drops if you commit for a longer period. Before committing to a long-term contract, consider that DSL service is new and prices will probably come down as competition increases. Check also for any early termination charges.

- **Is usage restricted?** Many ISPs include restrictions on the amount of data going through your DSL pipe per month per line. For example, there might be a 10-gigabyte limit for a lower-bandwidth DSL connection. Any data moving across your DSL line in excess of the limit costs extra, usually based on a price per megabyte.

- **Are servers allowed?** You may find an even harsher restriction that forbids you from running any servers or using an Ethernet router (or other Internet connection sharing solution) that supports NAT and DHCP to share the connection.

- **What are the payment terms?** Increasingly, ISPs are billing to credit cards to cut down on their accounts receivable overhead. Others bill on a monthly or annual basis, with the annual billing method offering a discount. Many ISPs also require a deposit to start the service.

How Good Is the DSL Provider or ISP?

You order DSL Internet service from an Internet service provider. Behind the ISP are CLECs and ILECs providing the DSL network service. The ISP can be heavily oversubscribing their DSL service and not providing good customer support. A DSL provider could be oversubscribing as well.

Unfortunately, helpful consumer information about specific ISPs and DSL providers is not readily available. DSL Reports (www.dslreports.com) is one of the few sites that provides helpful information for potential DSL customers. This site uses a database to present the results of thousands of DSL customer reviews.

Do You Plan to Connect Multiple Computers?

The high-speed, always-on nature of DSL service makes it ideal for LAN-to-Internet connections. The inherent benefit of building a LAN, beyond the sharing of local resources, is the capability to share the high-speed DSL connection across multiple computers. Even if you're using a single-user (dynamic IP address) DSL Internet access service, with a LAN you can share the service across multiple computers by adding a proxy server, Internet-connection-sharing software, or an Ethernet router.

Today, setting up the basic LAN plumbing for your office or home is easy and inexpensive. Building a network from the ground up involves adding network interface cards (NICs) to your computers and then connecting them to a network hub or a switch device. You can choose from Ethernet, Phoneline, and wireless networking products. See Chapter 11 for more information on networking.

Although sharing is something you might want to do with your Internet connection, check the ISP's terms and conditions for any legal restrictions on sharing the DSL service. Many consumer DSL offerings restrict you from using any kind of Internet-sharing solution.

What About Internet Security?

An always-on connection with a static IP address is wide open to hacker attacks that can rob you of your data, disrupt your Internet service with denial of services attacks, or destroy your computer operating system. Don't panic! Affordable security solutions for consumers, called personal firewalls, are available. For small businesses, more sophisticated firewall appliances can be used.

Glossary

• •

100BaseT: The Ethernet networking standard that supports a data transmission rate of 100 Mbps and is backward compatible to 10BaseT networks. 100BaseT is based on the IEEE 802.3u standard and is commonly referred to as Fast Ethernet.

10BaseT: The Ethernet networking standard that supports a data transmission rate of 10 Mbps. 10BaseT is based on the IEEE 802.3 specification.

2B1Q: Two Binary, One Quaternary. A line coding technique used in traditional telecommunications offerings including ISDN. The 2B1Q line coding is used for some DSL flavors, including SDSL, HDSL, and IDSL.

Access Point: The network hub device for a wireless network.

ADSL: Asymmetric Digital Subscriber Line. A DSL flavor that supports high-speed data communications speeds of up to 8 Mbps downstream and up to 640 Kbps upstream. ADSL can deliver simultaneous high-speed data and POTS voice service over the same telephone line. ADSL is the most widely deployed flavor of DSL by ILECs.

ADSL Forum: *See* DSL Forum.

ADSL Lite: Nickname for G.lite. A DSL flavor based on the new G.lite standard that supports 1.5 Mbps downstream and 348 Kbps upstream.

always on: Refers to a DSL Internet connection as a dedicated connection. There is no dial-up process to connect to the Internet because the computer is linked directly to the Internet.

ANSI: American National Standards Institute. The organization that defines standards, including network standards, for the United States.

ATM: Asynchronous Transfer Mode. A standard for high-speed digital backbone networks. ATM networks are widely used by telecommunications and large companies for backbone networks that consolidate data traffic from multiple feeders (such as DSL lines) and different types of media (voice, video, and data).

AWG: American Wire Gauge. A thickness measurement for copper wiring. The heavier the gauge, the lower the AWG number and the better the quality of the line in terms of supporting a longer distance for a DSL signal. Many local loops use 24 AWG or 26 AWG copper wires.

backbone: A major transmission path used for high-volume network-to-network connections. In DSL-to-Internet connections, a backbone network consolidates data traffic from the individual DSL lines into a backbone network for delivery to ISPs.

bandwidth: The amount of data that can flow through a given communications channel. The greater the bandwidth, the more data that can travel at one time.

best effort: Internet access service that doesn't have a Quality of Service (*QoS*) guarantee. Most DSL service is a best effort class of service, but QoS is emerging as a premium DSL service by some ISPs.

binary: A number system based on 2. The place columns of the number are based on powers of 2: 1, 2, 4, 8, 16, 32, 64, 128, 256, and so on. The binary to decimal conversions make up the IP addresses used on any TCP/IP network, such as 199.232.255.113.

bit: The single unit of data used in digital data communications. It takes 8 bits to make a byte, which is a measurement for computer data.

Bps: Bits per second. The unit of measurement for data transmission speed over a data communications link.

bridge: A device that connects two networks as a seamless single network using the same networking protocol. DSL modems are typically bridges. Bridges operate at the hardware layer and don't include IP routing capabilities.

bridge tap: An extension to a local loop generally used to attach a remote user to a central office switch without having to run a new pair of wires all the way back. Bridged taps are fine for POTS, but severely limit the speed of digital information flow on the link.

broadband: A term used to describe a high-capacity network that can carry several services on the same line, such as data, voice, and video. DSL is broadband.

byte: A unit of data consisting of 8 bits.

cable binder: A bundle of local loop wires that runs along telephone poles or underground from the CO.

CAP: Carrierless Amplitude Phase. A modulation transceiver technology used in ADSL systems.

CAT5: Category 5 unshielded twisted-pair wiring commonly used for 10BaseT and 100BaseT Ethernet networks and rated by the EIA/TIA.

channel: A path for digital transmission signals. Within digital services such as DSL, multiple channels can share the same pair of wires. Channels are created using multiplexers.

CIDR: Classless Internet Domain Routing. In response to the limitations of A, B, and C classes of IP addresses, InterNIC implemented CIDR (which is pronounced "cider"). CIDR allows IP addresses to be broken down into smaller subnets than the class C network, with 256 IP addresses. CIDR networks are described as slash x networks, where *x* represents the number of bits in the IP address range.

circuit: A path through a network from source to destination and back. In a circuit-switched network, this path uses a fixed route and a fixed amount of bandwidth for the duration of the connection between the two end points.

CLEC: Competitive Local Exchange Carrier. A competitor to ILECs offering telecommunication service. In the case of DSL service, the CLECs offer data communications service.

client: A program or a device that requests services from a server.

client/server: A style of computer networking that allows work to be distributed across powerful computers acting as servers and client computers. TCP/IP uses a client/server architecture.

CO: central office. A telephone company facility within which all local telephone lines terminate and which contains the equipment required to switch customer telecommunications traffic. For DSL service, DSLAM equipment is typically set up at the CO to support DSL service for lines terminating at the CO.

CPE: Customer Premises Equipment. A telecommunications term that refers to any equipment located at the customer's premises. DSL modems, bridges, and routers are CPE.

crosstalk: The interference induced on a signal on one line that is caused by the transfer of energy from a colocated line. Crosstalk is a factor in the delivery of different flavors of DSL service in the same cabling bundle.

CSMA/CD: Carrier Sense Multiple Access/Collision Detection. A network transmission scheme in which multiple network devices can transmit across the cable simultaneously. CSMA/CD is used as the basis of Ethernet networks.

data CLEC: A Competitive Local Exchange Carrier that focuses on IP data communications links and doesn't provide traditional voice telecommunications.

default gateway: The address that the IP uses if the destination address is not on the local subnet. The default gateway is usually the router's IP address.

demarcation point: The point at the customer premises where the line from the telephone company meets the premises wiring. From the demarcation point, the end user is responsible for the wiring.

DHCP: Dynamic Host Configuration Protocol. A protocol that allows IP addressing information to be dynamically assigned by a server to clients on an as-needed basis. IP addresses for a network are stored in a pool of available IP addresses, which are allocated when a computer on the network boots up. The DHCP server functionality is built into most DSL routers.

Dial-Up Networking (DUN): Used in Microsoft Windows 95, 98, NT, and 2000 for making PPP dial-up modem connections to the Internet. If you have a PCI or USB DSL modem that uses PPP over DSL, you need to use DUN.

DLC: Digital Loop Carrier. A telecommunications structure deployed wherever an ILEC needs more capacity. DLCs consist of a box containing line cards that concentrate individual lines within a given area and then send the traffic over a high-speed digital connection. DLCs are commonly deployed in new buildings, office parks, and residential subdivisions.

DMT: Discrete Multi-Tone. An ADSL modulation technique standardized by the ANSI T1E1.4 standard and used in ADSL systems.

DNS: Domain Name System. The name resolution service for IP addresses that provides the friendlier text-based addresses for Internet resources. DNS uses a distributed database containing FQDNs and addresses.

domain name: The user-friendly text name used instead of a numeric IP address for an Internet address.

domain name server: A program that converts an *FQDN* into its numeric IP address, and vice versa.

DNS service: Domain Name System service. The configuration of user-friendly text domain names to IP addresses by an ISP using the *Domain Name System.*

downstream: The direction of data flow on a data communications link that occurs from the network down to the user. In the case of Internet access, downstream is the capacity or speed of data flowing from the Internet to the end user's PC or LAN.

DSL: Digital Subscriber Line. The generic term that refers to the underlying technology inherent in all flavors of DSL, such as ADSL, SDSL, and HDSL.

DSL Forum: Previously know as the ADSL Forum. This industry organization is made up of computer and telecommunication companies that define DSL standards for submission to standards bodies. This group is responsible for accelerating ADSL technologies, products, and services as well as promoting the technology.

DSLAM: Digital Subscriber Line Access Multiplexer. The device typically housed at the CO that terminates all the DSL lines serviced by the CO. The DSLAM consolidates or concentrates all the data traffic coming in from individual DSL lines and passes them on to a backbone network for distribution to Internet service provider networks or corporate networks.

DSL bridge: A device that combines one or more networks into a single seamless network. Also called a DSL modem.

DSL modem: A common term used for a *DSL bridge*.

dynamic IP addressing: An IP address is assigned to the client for the current session or some other specified amount of time. This form of IP addressing is used by DSL services targeted at the consumer market and typically doesn't support any Domain Name System service.

EIA/TIA: Electronic Industries Association/Telecommunications Industry Association. An organization that provides standards for the data communications industry, including the cabling used for networking and telecommunications.

Ethernet: A LAN technology that uses CSMA/CD delivery and can run over different media (cabling). Most of today's Ethernet LANs use twisted-pair 10BaseT wiring that can support both standard Ethernet at 10 Mbps and Fast Ethernet at 100 Mbps.

Ethernet address: The unique hardware address that identifies any Ethernet device, including network interface cards (NICs), network printers, DSL bridges, and routers.

Fast Ethernet: The Ethernet-based networking protocol that supports up to 100-Mbps capacity. Commonly referred to as 100BaseT networking.

FCC: Federal Communications Commission. The United States government agency for regulating the telecommunications industry.

fiber optics: A technology in which light is used to transport large amounts of data using thin filaments of glass. *Fiber* is the common shorthand term for fiber-optics.

firewall: A security device (hardware or software) that controls access from the Internet to a local network by using identification information associated with TCP/IP packets to make a decision about whether to allow or deny access. This decision is based on a set of defined rules that describe which packets or sessions are allowed.

firmware: Instructions stored in memory that controls a device, such as a DSL modem or router. Because firmware can be overwritten to replace it, DSL equipment can be updated by installing new firmware into the device. This allows new features to be added to the DSL CPE without buying new equipment.

FQDN: Fully Qualified Domain Name. The full name of a host, including all subdomain and domain names, separated by dots. For example, david.support.angell.com is an FQDN.

fractional T-1: Any data transmission rate between 56 Kbps and 1.54 Mbps (which is the full T-1 rate). Fractional T-1 is simply a digital dedicated line that's not as fast and not as expensive as a T-1 line.

frame relay: A dedicated, public data networking service offered by telecommunication companies for LAN-to-LAN connections. Frame relay uses variable-length frames for packet-switching networks that efficiently handle bursty communications by quickly adjusting bandwidth to meet demands.

gateway: A functional device allows equipment with different protocols to communicate with each other. The gateway device can be embodied in a router or a computer.

G.lite: The new ITU standard that forms the basis of Universal ADSL, which supports 1.5 Mbps downstream and 384 Kbps upstream.

G.dmt: A standards-based form of ADSL that supports up to 8 Mbps downstream and 1.54 Mbps upstream.

hardware address: The physical address for the NIC, which is used by low-level hardware layers of the network, including DSL bridges. Also called the MAC address.

HDSL: High-bit-rate Digital Subscriber Line. The DSL service widely used for T-1 lines. HDSL uses four wires (two pairs) instead of the standard two wires used for other DSL flavors. HDSL supports symmetrical service at 1.54 Mbps but doesn't support POTS.

HDSL-2: High-bit-rate Digital Subscriber Line-2. The ITU has approved a new generation of HDSL that offers several enhancements over its predecessor. One of the most important is that HDSL-2 requires only a single twisted-pair local loop instead of the two pairs required for HDSL.

HomeRF: Home Radio Frequency. A wireless networking specification that uses the 2.4-GHz band.

host: A computer or any device connected to a TCP/IP network.

HomePNA: Home Phoneline Networking Alliance. The group that created the specification for Phoneline networking, which uses telephone wiring as network cabling. Network data traffic runs over the same telephone wiring used by telephones, fax machines, and analog modems, as well as ADSL/G.lite modems, without interrupting these services.

hub: A passive network device that repeats all data traffic to all ports. A hub is at the center of a LAN, and all networked devices, including computers, printers, and DSL bridges or routers, are connected to the hub through cables.

IDSL: ISDN Digital Subscriber Line. The always-on cousin of dial-up ISDN. IDSL delivers a symmetric 144 Kbps of bandwidth, which is 16 Kbps more than the dial-up version of ISDN. This 16-Kbps difference comes from the elimination of the two 8-Kbps channels used in ISDN for communicating with the PSTN switch. Unlike ISDN, IDSL doesn't support POTS.

IEEE: Institute of Electrical and Electronics Engineers. A worldwide engineering and standards-making body for the electronics industry. IEEE is the standards committee for LAN technologies and advises on standards for ANSI.

IEEE 802.3: The local area network protocol known as Ethernet. The 802.3 protocol forms the basis for 10-Mbps or 100-Mbps throughput that uses CSMA/CD. This allows LAN users to share the network cable, but only one station can use the cable at a time.

IEEE 802.11: The standard that forms the basis for wireless networks. It operates in the 2.4-GHz spectrum and supports a range of speeds from 1 Mbps to 11 Mbps.

IETF: Internet Engineering Task Force. The organization that provides the coordination of standards and specification development for TCP/IP networking. IETF is part of the IAB (Internet Architecture Board) and is responsible for research into Internet issues. RFCs (Request For Comments) document the IETF specifications.

ILEC: Incumbent Local Exchange Carrier. A new term that emerged from the Telecommunications Act of 1996 that describes the traditional local telephone companies, which control local telephone service (voice or data). Companies competing with these ILECs are called *CLEC*s.

IMAP4: Internet Message Access Protocol, Version 4. IMAP4 provides sophisticated client/server capabilities beyond the features of POP3. POP3 and IMAP4 don't interoperate, but many email servers and clients can support both protocols.

Internet address: The unique 32-bit numeric address, such as 199.232.255.113, used by a host on a TCP/IP network. The IP address consists of two parts: a network number and a host number.

interoperable: Two pieces of equipment are interoperable when they work together. Standards are designed to enable interoperability among different devices from different vendors.

intranet: A local network that uses TCP/IP and Web technologies as its networking protocol. Internal company information made available using a Web server and other TCP/IP applications. Many intranets are protected from exterior access by various security devices, such as routers, proxy servers, or firewalls.

IP: Internet Protocol. The connectionless network layer protocol that forms the networking functions of the TCP/IP suite. IP networking forms the basis of networking over the Internet and allows information to be transmitted across dissimilar networks.

IP address: Internet Protocol address. A 32-bit dotted decimal notation used to represent IP addresses. Each part of the address is a decimal number separated from other parts by a dot (.), such as 199.232.255.113.

IPSec: A virtual private networking protocol that is part of IPv6 but is widely used now in IPv4.

IPv4: The current version of IP addressing based on 32-bit IP addresses.

IPv6: The next generation of IP addressing based on 64-bit IP addresses and having a number of enhancements over IPv4, such as automatic IP address configuration and better security.

ISDN: Integrated Services Digital Network. An early member of DSL technology that can support up to 128-Kbps symmetrical service. It's routed through the ILEC's PSTN switches instead of through a DSLAM, as is the case for IDSL.

ISP: Internet service provider. Any company that provides Internet access service.

ITU: International Telecommunication Union. The ITU is an international body of member countries that defines recommendations and standards relating to international telecommunications.

IXC: Interexchange carrier. A long-distance telephone company.

Kbps: Kilobits per second. A measurement of digital bandwidth where one Kbps equals one thousand (actually 1,024) bits per second.

L2TP: Layer 2 Tunneling Protocol. An IETF protocol used for virtual private networking.

LAN: Local Area Network. A data network that connects computers in an area usually within the confines of a building or floors within a building. A LAN enables users to share information and network resources, such as a printer or DSL CPE. Ethernet forms the basis of most local area networks.

last mile: The telephone line between a local telephone company switching facility and the customer premises. Also called the *local loop.*

latency: A measure of the delay between the sending of a packet at the originating end of a connection and the reception of that packet at the destination end.

layer: In the OSI network reference model, each layer performs a certain task to move the data from the sender to the receiver. Protocols within the layers define the tasks for the networks.

LEC: Local Exchange Carrier. The company that provides local voice and data services. Both ILECs and CLECs are local exchange carriers.

loading coil: A metallic, doughnut-shaped device used on local loops to extend their reach. Loading coils severely limit the bandwidth in digital communications.

local loop: A generic term for the connection between the customer's premises and the telephone company's serving central office. The local loop is the pair of copper wires that connects the end user to the central office, which is the gateway to the telecommunications network.

MAC address: Media Access Control address. The 48-bit defined number built into any Ethernet device connected to a LAN. This unique hardware address is represented as six octets, separated by colons, such as C0:3C:4E:00:10:8F. Bridges work at the MAC address level.

Mbps: Mega bits per second. A measurement of digital bandwidth where one Mbps equals just over one million bits per second.

MDF: Main Distribution Frame. The point where all local loops are terminated at a CO.

MPOE: Minimum Point of Entry. The place where phone lines first enter a customer's facility. The MPOE can be a network interface device or an inside wiring closet.

multiplexer: Any one of a number of common devices used to combine multiple telecommunications circuits into channels. DSL lines coming into the CO are multiplexed to be carried over trunk lines.

MVL: Multiple Virtual Lines. A DSL technology developed by Paradyne. MVL transforms a single copper loop into multiple virtual lines to support multiple independent services over the same line simultaneously.

NAT: Network Address Translation. An Internet standard that allows your local network to use private IP addresses, which are not recognized on the Internet. The IP address used for the router is the only routable IP address. The computers behind the NAT can access the Internet through the router, but Internet users can't access the computers behind the router.

NDIS: Network Driver Interface Specification. Developed by Microsoft to provide a common set of rules for network adapters to interface with operating systems.

NIC: Network interface card. The hardware that forms the interface between the computer (or other network device) and not only the data communications network for the LAN but also the IP connection through a DSL Ethernet bridge or router.

NID: Network interface device. A device that terminates a copper pair from the serving central office at the user's destination. The NID is typically a small box installed on the exterior premises of the destination.

NNTP: Network News Transport Protocol. The protocol that governs the transmission of network news, a threaded messaging system for posting messages to form newsgroup discussions.

NSP: Network service provider. Any company that provides network services to subscribers.

OSI: Open Systems Interconnection. An internationally accepted model of data communication protocols developed by OSI and ITU. The OSI Reference Model has seven layers of protocols used for networking.

packet: A fixed- or variable-sized unit of information that can be sent across a packet-switching network. A packet typically contains addressing information, error checking, and user information in addition to application data. IP packets vary in size.

packet CLEC: A Competitive Local Exchange Carrier that focuses on providing data communication services instead of voice services. The term was coined by Covad Communications to describe their data-centric services.

packet filter: The capability to search a packet to determine its destination and then route or block it accordingly. Routers perform this function to route TCP/IP data traffic.

packet-switched network: A network that does not establish a dedicated path through the network for the duration of a session but instead transmits data in units called packets in a connectionless manner. Data streams are broken into packets at the front end of a transmission.

packet switching: A data transmission method in which data is transferred by packets, or blocks of data. Packets are sent using a store-and-forward method across nodes in a network.

PC Card: The credit-card-size adapter cards used in notebooks. A DSL modem for a notebook can be a PC Card.

PCI: Peripheral Component Interconnect. A specification introduced by Intel that defines a local bus system that allows up to ten PCI-compliant expansion cards in a PC. PCI is a de facto bus standard for today's PCs that has replaced the ISA (Industry Standard Architecture) bus.

personal firewall: A software program that runs on your computer to provide protection from Internet intruders.

Phoneline network: A networking technology based on the HomePNA specification that uses telephone wiring as network cabling. Network data traffic runs over telephone wiring used by telephones, fax machines, analog modems, and ADSL/G.lite modems without interrupting these services. Phoneline NICs are installed in PCs in the same way Ethernet adapters are but they connect to a telephone line instead of 10BaseT cabling. Phoneline networking uses the daisy chain topology that links computers to each other without the use of a hub.

PnP: Plug-and-Play. A system for simplifying installation of hardware devices on a Microsoft Windows computer. PnP automates hardware recognition, driver installation, and system management.

POP3: Post Office Protocol, Version 3. The latest version of the Post Office Protocol, POP3 provides basic client/server features for handling email. Most email client programs support POP3.

POTS: Plain Old Telephone Service. A historical term for basic telephone voice service over two-wire copper loop and out to the PSTN.

PPP: Point-to-Point Protocol. A communications protocol that allows a computer using TCP/IP to connect directly to the Internet through a dial-up connection. In Microsoft Windows, this type of connection is set up and controlled using Dial-Up Networking (DUN).

PPPoA: Point-to-Point Protocol over Asynchronous Transfer Mode. ATM is a high-speed switching technique used to transmit high volumes of voice, data, and video traffic. Using PPP over ATM enables TCP/IP traffic to be carried over an ATM network all the way down to a computer without being translated. This configuration requires an ATM adapter card in each computer that connects to an ATM ADSL bridge or router.

PPPoE: Point-to-Point Protocol over Ethernet. A standard that enables dial-up networking capabilities over Ethernet. PPPoE is a software driver that works with a NIC to create a dial-up session through the NIC and the LAN out through the DSL bridge or router. The significance of PPP over Ethernet has to do with making DSL service installation easier for users and ISPs, as well as enabling users to access multiple network services from the same DSL connection.

PPTP: Point-to-Point Tunneling Protocol is the VPN client software solution included with Microsoft Windows 95, 98, NT, and 2000. The PPTP server is included in Windows NT server as part of Remote Access Service (RAS) and Routing and Remote Access Server (RRAS). A PPTP server is built into some DSL routers.

protocol: A set of rules that defines how different systems interoperate.

PSTN: Public Switched Telephone Network. The network that provides global telephone service.

PUC: Public Utility Commission. A United States government agency, usually at the state level, that regulates telecommunication companies and other utilities.

QoS: Quality of Service. A definition of a given level of service for voice or data communication services by the provider.

RBOCs: Regional Bell Operating Companies. The seven original regional Bell operating companies that provided local telephone service and were formed as a result of the AT&T divestiture. They are Ameritech, Bell Atlantic, Bell South, NYNEX, Pacific Bell, Southwestern Bell, and US WEST.

RJ-11: A standard modular connector (jack or plug) that supports two pairs of wires (four wires). RJ-11 is commonly used for most PSTN CPE (telephones, fax machines, and modems).

RJ-45: A standard modular connector that can support up to four pairs of wires (eight wires). RJ-45 connectors are used with Category 5 cabling used with 10BaseT or 100BaseT cabling. This cabling is also used in business locations with more sophisticated voice communication systems.

router: A device that routes data between networks through IP addressing information contained in the header of the IP packet. A router forwards packets to other routers until the packets reach their destination. Routers form the basis of IP networking.

SDSL: Symmetrical Digital Subscriber Line. A member of the DSL family that is widely deployed by CLECs. SDSL supports symmetrical service at 160 Kbps to 2.3 Mbps but does not support POTS connections. SDSL reaches up to 18,000 feet from the CO.

server: A host that makes an application or a service available to other hosts, typically clients. For example, a Web server relays information to Web browser clients.

SMTP: Simple Mail Transfer Protocol. SMTP is the protocol for Internet email that transfers email messages among computers. SMTP uses a store-and-forward system to move email messages to their final destination.

splitter: A device used to separate POTS service from the ADSL data service at a customer's premises. The CO side of the DSL connection also has a POTS splitter.

SSL: Secure Sockets Layer. SSL version 2 provides security by allowing applications to encrypt data that goes from a client, such as Web browser, to a matching server. (Encrypting your data means converting it to a secret code.) SSL version 3 allows the server to authenticate that the client is who it says it is.

standard: A set of technical specifications used to establish uniformity in software, hardware, and data communications.

static IP addressing: An assigned IP address used to connect to a TCP/IP network. The IP address stays with the specific host or network device. Typically used with routable public IP addresses so that a particular host can be reached by its assigned static IP address and any domain name associated with the IP address.

STP: Shielded twisted pair. A shielded form of the twisted-pair wiring used for 10BaseT and 100BaseT LANs. STP has a foil or wire braid wrapped around the individual wires to provide better protection against electromagnetic interference. STP uses different connectors than UTP, is more expensive than UTP, and requires careful grounding to work properly.

subnet: A portion of a network. Each subnet within a network shares a common network address and is uniquely identified by a subnetwork number.

subnet mask: A 32-bit number used to separate the network and host sections of an IP address. A subnet mask subdivides an IP network into smaller pieces. An example of a subnet mask address might be 255.255.255.248 for an 8 IP address network.

T-1: A North American standard for communicating at 1.54 Mbps. A T-1 line has the capacity for 24 voice and data channels at 64 Kbps each.

T-3: A North American standard for communicating at speeds of 44 Mbps. A T-3 line has 672 channels for voice and data at 64 Kbps each.

TCP: Transmission Control Protocol. One of two principal components of the TCP/IP protocol suite. TCP puts data into packets and provides packet delivery across the network, ensuring that packets are not lost in transmission and arrive in order.

TCP/IP: Transmission Control Protocol/Internet Protocol. TCP/IP is the suite of protocols that define the basis of the Internet. It provides communications across interconnected networks between computers with diverse hardware and operating systems.

TCP/IP stack: The software that allows a computer to communicate through TCP/IP. Stack refers to the fact that five layers of protocols operate on a TCP/IP network.

Telecommunications Act of 1996: Legislation passed by the United States Congress that has opened up local telecommunications to competition. The goal is to give consumers more choices at lower costs. Although by no means a perfect piece of legislation, it's already bearing fruit in the deployment of DSL.

telephony: The science of transmitting voice, data, and video over a distance greater than you can transmit by shouting.

telnet: A terminal-emulation protocol that allows you to access computers and network devices through TCP/IP.

twisted pair: A cable comprised of pairs of wires twisted around each other to help cancel out interference. This is the common form of copper cabling used for telephony and data communications.

UART: Universal Asynchronous Receiver/Transceiver. The older serial port architecture for data communications that is limited to 115-Kbps capacity. UART is being replaced with *USB* (Universal Serial Bus).

upstream: The direction of data traffic from a computer to the Internet. The faster the upstream speed, the faster data can move from your local network or PC to the Internet. If you run a Web server or any type of TCP/IP server, it will affect your upstream capacity.

USB: Universal Serial Bus. A new data communications port installed on most newer PCs to replace the UART serial port. USB ports are easier to use for plugging in peripherals and support data communication speeds up to 12 Mbps.

UTP: Unshielded twisted pair. Cabling used for 10BaseT and 100BaseT LANs. UTP consists of pairs of copper wires twisted around each other and covered by plastic insulation. UTP is by far the most popular cabling used for LANs.

VDSL: Very-high-bit-rate Digital Subscriber Line. An ultra-high-speed DSL flavor that can deliver data communications at speeds up to 52 Mbps for short distances of up to 4,000 feet.

VoDSL: Voice over DSL. A hybrid voice communication system that enables digital voice communications over a DSL network and then passes the voice to the PSTN.

VoIP: Voice over IP. Forms the basis of PC-to-PC voice communications over the Internet. Products such as Microsoft NetMeeting, Net2Phone, and Internet Phone allow you to do VoIP communications using the sound card, speakers, and microphone (or headset) connected to your computer.

VPN: Virtual private network. A way that private data can safely pass over a public network, such as the Internet. The data traveling between the two hosts is encrypted for privacy, and other security features are included to provide a secure direct connection over the Internet.

WAN: Wide area network. A data network typically extending a LAN outside a building over a data communications link to another network in another location. A WAN typically uses common carrier lines. The jump between a LAN and a WAN is made through a bridge or a router.

Web hosting: A service performed by an ISP or a Web hosting service that operates all the Web server infrastructure for you.

wireless network: A new networking technology that uses radio waves. Second-generation wireless networking products support 11 Mbps (about the same speed as standard Ethernet).

xDSL: A generic term used to refer to the entire family of DSL technologies. The *x* is a placeholder for *A* in ADSL, *S* in SDSL, and so on.

Index

M

N

X

Z

Notes

Notes

Notes

Notes

Notes

Notes

From PCs
to Personal Finance,
We Make it Fun and Easy!

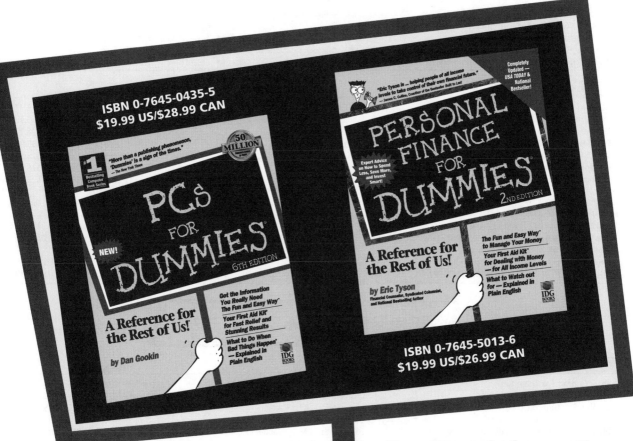

For more information,
or to order, please
call 800.762.2974.

www.idgbooks.com
www.dummies.com

Dummies Books™
Bestsellers on Every Topic!

TECHNOLOGY TITLES

INTERNET

America Online® For Dummies®, 5th Edition	John Kaufeld	0-7645-0502-5	$19.99 US/$26.99 CAN
E-Mail For Dummies®, 2nd Edition	John R. Levine, Carol Baroudi, Margaret Levine Young, & Arnold Reinhold	0-7645-0131-3	$24.99 US/$34.99 CAN
Genealogy Online For Dummies®	Matthew L. Helm & April Leah Helm	0-7645-0377-4	$24.99 US/$35.99 CAN
Internet Directory For Dummies®, 2nd Edition	Brad Hill	0-7645-0436-3	$24.99 US/$35.99 CAN
The Internet For Dummies®, 6th Edition	John R. Levine, Carol Baroudi, & Margaret Levine Young	0-7645-0506-8	$19.99 US/$28.99 CAN
Investing Online For Dummies®, 2nd Edition	Kathleen Sindell, Ph.D.	0-7645-0509-2	$24.99 US/$35.99 CAN
World Wide Web Searching For Dummies®, 2nd Edition	Brad Hill	0-7645-0264-6	$24.99 US/$34.99 CAN

OPERATING SYSTEMS

DOS For Dummies®, 3rd Edition	Dan Gookin	0-7645-0361-8	$19.99 US/$28.99 CAN
LINUX® For Dummies®, 2nd Edition	John Hall, Craig Witherspoon, & Coletta Witherspoon	0-7645-0421-5	$24.99 US/$35.99 CAN
Mac® OS 8 For Dummies®	Bob LeVitus	0-7645-0271-9	$19.99 US/$26.99 CAN
Small Business Windows® 98 For Dummies®	Stephen Nelson	0-7645-0425-8	$24.99 US/$35.99 CAN
UNIX® For Dummies®, 4th Edition	John R. Levine & Margaret Levine Young	0-7645-0419-3	$19.99 US/$28.99 CAN
Windows® 95 For Dummies®, 2nd Edition	Andy Rathbone	0-7645-0180-1	$19.99 US/$26.99 CAN
Windows® 98 For Dummies®	Andy Rathbone	0-7645-0261-1	$19.99 US/$28.99 CAN

PC/GENERAL COMPUTING

Buying a Computer For Dummies®	Dan Gookin	0-7645-0313-8	$19.99 US/$28.99 CAN
Illustrated Computer Dictionary For Dummies®, 3rd Edition	Dan Gookin & Sandra Hardin Gookin	0-7645-0143-7	$19.99 US/$26.99 CAN
Modems For Dummies®, 3rd Edition	Tina Rathbone	0-7645-0069-4	$19.99 US/$26.99 CAN
Small Business Computing For Dummies®	Brian Underdahl	0-7645-0287-5	$24.99 US/$35.99 CAN
Upgrading & Fixing PCs For Dummies®, 4th Edition	Andy Rathbone	0-7645-0418-5	$19.99 US/$28.99CAN

GENERAL INTEREST TITLES

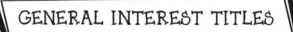

FOOD & BEVERAGE/ENTERTAINING

Entertaining For Dummies®	Suzanne Williamson with Linda Smith	0-7645-5027-6	$19.99 US/$26.99 CAN
Gourmet Cooking For Dummies®	Charlie Trotter	0-7645-5029-2	$19.99 US/$26.99 CAN
Grilling For Dummies®	Marie Rama & John Mariani	0-7645-5076-4	$19.99 US/$26.99 CAN
Italian Cooking For Dummies®	Cesare Casella & Jack Bishop	0-7645-5098-5	$19.99 US/$26.99 CAN
Wine For Dummies®, 2nd Edition	Ed McCarthy & Mary Ewing-Mulligan	0-7645-5114-0	$19.99 US/$26.99 CAN

SPORTS

Baseball For Dummies®	Joe Morgan with Richard Lally	0-7645-5085-3	$19.99 US/$26.99 CAN
Fly Fishing For Dummies®	Peter Kaminsky	0-7645-5073-X	$19.99 US/$26.99 CAN
Football For Dummies®	Howie Long with John Czarnecki	0-7645-5054-3	$19.99 US/$26.99 CAN
Hockey For Dummies®	John Davidson with John Steinbreder	0-7645-5045-4	$19.99 US/$26.99 CAN
Tennis For Dummies®	Patrick McEnroe with Peter Bodo	0-7645-5087-X	$19.99 US/$26.99 CAN

HOME & GARDEN

Decks & Patios For Dummies®	Robert J. Beckstrom & National Gardening Association	0-7645-5075-6	$16.99 US/$24.99 CAN
Flowering Bulbs For Dummies®	Judy Glattstein & National Gardening Association	0-7645-5103-5	$16.99 US/$24.99 CAN
Home Improvement For Dummies®	Gene & Katie Hamilton & the Editors of HouseNet, Inc.	0-7645-5005-5	$19.99 US/$26.99 CAN
Lawn Care For Dummies®	Lance Walheim & National Gardening Association	0-7645-5077-2	$16.99 US/$24.99 CAN

IDG BOOKS WORLDWIDE

*For more information, or to order,
call (800)762-2974*

BESTSELLI BOOK SER

Dummies Books™
Bestsellers on Every Topic!

TECHNOLOGY TITLES

SUITES

Title	Author	ISBN	Price
Microsoft® Office 2000 For Windows® For Dummies®	Wallace Wang & Roger C. Parker	0-7645-0452-5	$19.99 US/$28.99 CAN
Microsoft® Office 2000 For Windows® For Dummies®, Quick Reference	Doug Lowe & Bjoern Hartsfvang	0-7645-0453-3	$12.99 US/$19.99 CAN
Microsoft® Office 4 For Windows® For Dummies®	Roger C. Parker	1-56884-183-3	$19.95 US/$26.95 CAN
Microsoft® Office 97 For Windows® For Dummies®	Wallace Wang & Roger C. Parker	0-7645-0050-3	$19.99 US/$26.99 CAN
Microsoft® Office 97 For Windows® For Dummies®, Quick Reference	Doug Lowe	0-7645-0062-7	$12.99 US/$17.99 CAN
Microsoft® Office 98 For Macs® For Dummies®	Tom Negrino	0-7645-0229-8	$19.99 US/$28.99 CAN

WORD PROCESSING

Title	Author	ISBN	Price
Word 2000 For Windows® For Dummies®, Quick Reference	Peter Weverka	0-7645-0449-5	$12.99 US/$19.99 CAN
Corel® WordPerfect® 8 For Windows® For Dummies®	Margaret Levine Young, David Kay, & Jordan Young	0-7645-0186-0	$19.99 US/$26.99 CAN
Word 2000 For Windows® For Dummies®	Dan Gookin	0-7645-0448-7	$19.99 US/$28.99 CAN
Word For Windows® 95 For Dummies®	Dan Gookin	1-56884-932-X	$19.99 US/$26.99 CAN
Word 97 For Windows® For Dummies®	Dan Gookin	0-7645-0052-X	$19.99 US/$26.99 CAN
WordPerfect® 6.1 For Windows® For Dummies®, Quick Reference, 2nd Edition	Margaret Levine Young & David Kay	1-56884-966-4	$9.99 US/$12.99 CAN
WordPerfect® 7 For Windows® 95 For Dummies®	Margaret Levine Young & David Kay	1-56884-949-4	$19.99 US/$26.99 CAN
Word Pro® for Windows® 95 For Dummies®	Jim Meade	1-56884-232-5	$19.99 US/$26.99 CAN

SPREADSHEET/FINANCE/PROJECT MANAGEMENT

Title	Author	ISBN	Price
Excel For Windows® 95 For Dummies®	Greg Harvey	1-56884-930-3	$19.99 US/$26.99 CAN
Excel 2000 For Windows® For Dummies®	Greg Harvey	0-7645-0446-0	$19.99 US/$28.99 CAN
Excel 2000 For Windows® For Dummies® Quick Reference	John Walkenbach	0-7645-0447-9	$12.99 US/$19.99 CAN
Microsoft® Money 98 For Dummies®	Peter Weverka	0-7645-0295-6	$24.99 US/$34.99 CAN
Microsoft® Money 99 For Dummies®	Peter Weverka	0-7645-0433-9	$19.99 US/$28.99 CAN
Microsoft® Project 98 For Dummies®	Martin Doucette	0-7645-0321-9	$24.99 US/$34.99 CAN
MORE Excel 97 For Windows® For Dummies®	Greg Harvey	0-7645-0138-0	$22.99 US/$32.99 CAN
Quicken® 98 For Windows® For Dummies®	Stephen L. Nelson	0-7645-0243-3	$19.99 US/$26.99 CAN

GENERAL INTEREST TITLES

EDUCATION & TEST PREPARATION

Title	Author	ISBN	Price
The ACT For Dummies®	Suzee Vlk	1-56884-387-9	$14.99 US/$21.99 CAN
College Financial Aid For Dummies®	Dr. Herm Davis & Joyce Lain Kennedy	0-7645-5049-7	$19.99 US/$26.99 CAN
College Planning For Dummies®, 2nd Edition	Pat Ordovensky	0-7645-5048-9	$19.99 US/$26.99 CAN
Everyday Math For Dummies®	Charles Seiter, Ph.D.	1-56884-248-1	$14.99 US/$22.99 CAN
The GMAT® For Dummies®, 3rd Edition	Suzee Vlk	0-7645-5082-9	$16.99 US/$24.99 CAN
The GRE® For Dummies®, 3rd Edition	Suzee Vlk	0-7645-5083-7	$16.99 US/$24.99 CAN
Politics For Dummies®	Ann DeLaney	1-56884-381-X	$19.99 US/$26.99 CAN
The SAT I For Dummies®, 3rd Edition	Suzee Vlk	0-7645-5044-6	$14.99 US/$21.99 CAN

CAREERS

Title	Author	ISBN	Price
Cover Letters For Dummies®	Joyce Lain Kennedy	1-56884-395-X	$12.99 US/$17.99 CAN
Cool Careers For Dummies®	Marty Nemko, Paul Edwards, & Sarah Edwards	0-7645-5095-0	$16.99 US/$24.99 CAN
Job Hunting For Dummies®	Max Messmer	1-56884-388-7	$16.99 US/$24.99 CAN
Job Interviews For Dummies®	Joyce Lain Kennedy	1-56884-859-5	$12.99 US/$17.99 CAN
Resumes For Dummies®, 2nd Edition	Joyce Lain Kennedy	0-7645-5113-2	$12.99 US/$17.99 CAN

IDG BOOKS

For more information, or to order, call (800)762-2974

Dummies Books™ Bestsellers on Every Topic!

TECHNOLOGY TITLES

WEB DESIGN & PUBLISHING

Creating Web Pages For Dummies®, 4th Edition	Bud Smith & Arthur Bebak	0-7645-0504-1	$24.99 US/$34.99 CAN
FrontPage® 98 For Dummies®	Asha Dornfest	0-7645-0270-0	$24.99 US/$34.99 CAN
HTML 4 For Dummies®	Ed Tittel & Stephen Nelson James	0-7645-0331-6	$29.99 US/$42.99 CAN
Java™ For Dummies®, 2nd Edition	Aaron E. Walsh	0-7645-0140-2	$24.99 US/$34.99 CAN
PageMill™ 2 For Dummies®	Deke McClelland & John San Filippo	0-7645-0028-7	$24.99 US/$34.99 CAN

DESKTOP PUBLISHING GRAPHICS/MULTIMEDIA

CorelDRAW™ 8 For Dummies®	Deke McClelland	0-7645-0317-0	$19.99 US/$26.99 CAN
Desktop Publishing and Design For Dummies®	Roger C. Parker	1-56884-234-1	$19.99 US/$26.99 CAN
Digital Photography For Dummies®, 2nd Edition	Julie Adair King	0-7645-0431-2	$19.99 US/$28.99 CAN
Microsoft® Publisher 97 For Dummies®	Barry Sosinsky, Christopher Benz & Jim McCarter	0-7645-0148-8	$19.99 US/$26.99 CAN
Microsoft® Publisher 98 For Dummies®	Jim McCarter	0-7645-0395-2	$19.99 US/$28.99 CAN

MACINTOSH

Macs® For Dummies®, 6th Edition	David Pogue	0-7645-0398-7	$19.99 US/$28.99 CAN
Macs® For Teachers™, 3rd Edition	Michelle Robinette	0-7645-0226-3	$24.99 US/$34.99 CAN
The iMac For Dummies	David Pogue	0-7645-0495-9	$19.99 US/$26.99 CAN

GENERAL INTEREST TITLES

BUSINESS & PERSONAL FINANCE

Accounting For Dummies®	John A. Tracy, CPA	0-7645-5014-4	$19.99 US/$26.99 CAN
Business Plans For Dummies®	Paul Tiffany, Ph.D. & Steven D. Peterson, Ph.D.	1-56884-868-4	$19.99 US/$26.99 CAN
Consulting For Dummies®	Bob Nelson & Peter Economy	0-7645-5034-9	$19.99 US/$26.99 CAN
Customer Service For Dummies®	Karen Leland & Keith Bailey	1-56884-391-7	$19.99 US/$26.99 CAN
Home Buying For Dummies®	Eric Tyson, MBA & Ray Brown	1-56884-385-2	$16.99 US/$24.99 CAN
House Selling For Dummies®	Eric Tyson, MBA & Ray Brown	0-7645-5038-1	$16.99 US/$24.99 CAN
Investing For Dummies®	Eric Tyson, MBA	1-56884-393-3	$19.99 US/$26.99 CAN
Law For Dummies®	John Ventura	1-56884-860-9	$19.99 US/$26.99 CAN
Managing For Dummies®	Bob Nelson & Peter Economy	1-56884-858-7	$19.99 US/$26.99 CAN
Marketing For Dummies®	Alexander Hiam	1-56884-699-1	$19.99 US/$26.99 CAN
Mutual Funds For Dummies®, 2nd Edition	Eric Tyson, MBA	0-7645-5112-4	$19.99 US/$26.99 CAN
Negotiating For Dummies®	Michael C. Donaldson & Mimi Donaldson	1-56884-867-6	$19.99 US/$26.99 CAN
Personal Finance For Dummies®, 2nd Edition	Eric Tyson, MBA	0-7645-5013-6	$19.99 US/$26.99 CAN
Personal Finance For Dummies® For Canadians	Eric Tyson, MBA & Tony Martin	1-56884-378-X	$18.99 US/$24.99 CAN
Sales Closing For Dummies®	Tom Hopkins	0-7645-5063-2	$14.99 US/$21.99 CAN
Sales Prospecting For Dummies®	Tom Hopkins	0-7645-5066-7	$14.99 US/$21.99 CAN
Selling For Dummies®	Tom Hopkins	1-56884-389-5	$16.99 US/$24.99 CAN
Small Business For Dummies®	Eric Tyson, MBA & Jim Schell	0-7645-5094-2	$19.99 US/$26.99 CAN
Small Business Kit For Dummies®	Richard D. Harroch	0-7645-5093-4	$24.99 US/$34.99 CAN
Successful Presentations For Dummies®	Malcolm Kushner	1-56884-392-5	$16.99 US/$24.99 CAN
Time Management For Dummies®	Jeffrey J. Mayer	1-56884-360-7	$16.99 US/$24.99 CAN

AUTOMOTIVE

Auto Repair For Dummies®	Deanna Sclar	0-7645-5089-6	$19.99 US/$26.99 CAN
Buying A Car For Dummies®	Deanna Sclar	0-7645-5091-8	$16.99 US/$24.99 CAN
Car Care For Dummies®: The Glove Compartment Guide	Deanna Sclar	0-7645-5090-X	$9.99 US/$13.99 CAN

IDG BOOKS WORLDWIDE®

For more information, or to order, call (800)762-2974

BESTSELLING BOOK SERIE

Dummies Books™
Bestsellers on Every Topic!

TECHNOLOGY TITLES

DATABASE

Access 2000 For Windows® For Dummies®	John Kaufeld	0-7645-0444-4	$19.99 US/$28.99 CAN
Access 97 For Windows® For Dummies®	John Kaufeld	0-7645-0048-1	$19.99 US/$26.99 CAN
Approach® 97 For Windows® For Dummies®	Deborah S. Ray & Eric J. Ray	0-7645-0001-5	$19.99 US/$26.99 CAN
Crystal Reports 7 For Dummies®	Douglas J. Wolf	0-7645-0548-3	$24.99 US/$34.99 CAN
Data Warehousing For Dummies®	Alan R. Simon	0-7645-0170-4	$24.99 US/$34.99 CAN
FileMaker® Pro 4 For Dummies®	Tom Maremaa	0-7645-0210-7	$19.99 US/$26.99 CAN
Intranet & Web Databases For Dummies®	Paul Litwin	0-7645-0221-2	$29.99 US/$42.99 CAN

NETWORKING

Building An Intranet For Dummies®	John Fronckowiak	0-7645-0276-X	$29.99 US/$42.99 CAN
cc: Mail™ For Dummies®	Victor R. Garza	0-7645-0055-4	$19.99 US/$26.99 CAN
Client/Server Computing For Dummies®, 2ⁿᵈ Edition	Doug Lowe	0-7645-0066-X	$24.99 US/$34.99 CAN
Lotus Notes® Release 4 For Dummies®	Stephen Londergan & Pat Freeland	1-56884-934-6	$19.99 US/$26.99 CAN
Networking For Dummies®, 4ᵗʰ Edition	Doug Lowe	0-7645-0498-3	$19.99 US/$28.99 CAN
Upgrading & Fixing Networks For Dummies®	Bill Camarda	0-7645-0347-2	$29.99 US/$42.99 CAN
Windows NT® Networking For Dummies®	Ed Tittel, Mary Madden, & Earl Follis	0-7645-0015-5	$24.99 US/$34.99 CAN

GENERAL INTEREST TITLES

THE ARTS

Blues For Dummies®	Lonnie Brooks, Cub Koda, & Wayne Baker Brooks	0-7645-5080-2	$24.99 US/$34.99 CAN
Classical Music For Dummies®	David Pogue & Scott Speck	0-7645-5009-8	$24.99 US/$34.99 CAN
Guitar For Dummies®	Mark Phillips & Jon Chappell of Cherry Lane Music	0-7645-5106-X	$24.99 US/$34.99 CAN
Jazz For Dummies®	Dirk Sutro	0-7645-5081-0	$24.99 US/$34.99 CAN
Opera For Dummies®	David Pogue & Scott Speck	0-7645-5010-1	$24.99 US/$34.99 CAN
Piano For Dummies®	Blake Neely of Cherry Lane Music	0-7645-5105-1	$24.99 US/$34.99 CAN

HEALTH & FITNESS

Beauty Secrets For Dummies®	Stephanie Seymour	0-7645-5078-0	$19.99 US/$26.99 CAN
Fitness For Dummies®	Suzanne Schlosberg & Liz Neporent, M.A.	1-56884-866-8	$19.99 US/$26.99 CAN
Nutrition For Dummies®	Carol Ann Rinzler	0-7645-5032-2	$19.99 US/$26.99 CAN
Sex For Dummies®	Dr. Ruth K. Westheimer	1-56884-384-4	$16.99 US/$24.99 CAN
Weight Training For Dummies®	Liz Neporent, M.A. & Suzanne Schlosberg	0-7645-5036-5	$19.99 US/$26.99 CAN

LIFESTYLE/SELF-HELP

Dating For Dummies®	Dr. Joy Browne	0-7645-5072-1	$19.99 US/$26.99 CAN
Parenting For Dummies®	Sandra H. Gookin	1-56884-383-6	$16.99 US/$24.99 CAN
Success For Dummies®	Zig Ziglar	0-7645-5061-6	$19.99 US/$26.99 CAN
Weddings For Dummies®	Marcy Blum & Laura Fisher Kaiser	0-7645-5055-1	$19.99 US/$26.99 CAN

For more information, or to order, call (800)762-2974

Discover Dummies Online!

The Dummies Web Site is your fun and friendly online resource for the latest information about *For Dummies*® books and your favorite topics. The Web site is the place to communicate with us, exchange ideas with other *For Dummies* readers, chat with authors, and have fun!

Ten Fun and Useful Things You Can Do at www.dummies.com

1. Win free *For Dummies* books and more!
2. Register your book and be entered in a prize drawing.
3. Meet your favorite authors through the IDG Books Worldwide Author Chat Series.
4. Exchange helpful information with other *For Dummies* readers.
5. Discover other great *For Dummies* books you must have!
6. Purchase Dummieswear® exclusively from our Web site.
7. Buy *For Dummies* books online.
8. Talk to us. Make comments, ask questions, get answers!
9. Download free software.
10. Find additional useful resources from authors.

Link directly to these ten fun and useful things at
http://www.dummies.com/10useful

For other technology titles from IDG Books Worldwide, go to
www.idgbooks.com

Not on the Web yet? It's easy to get started with *Dummies 101*®: *The Internet For Windows*® *98* or *The Internet For Dummies*® at local retailers everywhere.

Find other *For Dummies* books on these topics:
Business • Career • Databases • Food & Beverage • Games • Gardening • Graphics • Hardware
Health & Fitness • Internet and the World Wide Web • Networking • Office Suites
Operating Systems • Personal Finance • Pets • Programming • Recreation • Sports
Spreadsheets • Teacher Resources • Test Prep • Word Processing

IDG BOOKS WORLDWIDE BOOK REGISTRATION

We want to hear from you!

Visit **http://my2cents.dummies.com** to register this book and tell us how you liked it!

✔ Get entered in our monthly prize giveaway.

✔ Give us feedback about this book — tell us what you like best, what you like least, or maybe what you'd like to ask the author and us to change!

✔ Let us know any other *For Dummies*® topics that interest you.

Your feedback helps us determine what books to publish, tells us what coverage to add as we revise our books, and lets us know whether we're meeting your needs as a *For Dummies* reader. You're our most valuable resource, and what you have to say is important to us!

Not on the Web yet? It's easy to get started with *Dummies 101*®: *The Internet For Windows*® *98* or *The Internet For Dummies*® at local retailers everywhere.

Or let us know what you think by sending us a letter at the following address:

For Dummies Book Registration
Dummies Press
10475 Crosspoint Blvd.
Indianapolis, IN 46256

BESTSELLING
BOOK SERIES